HIDDEN

ENERGY

Tesla-inspired Inventors and a Mindful Path to Energy Abundance

JEANE MANNING & SUSAN MANEWICH

 FriesenPress

Suite 300 - 990 Fort St
Victoria, BC, V8V 3K2
Canada

www.friesenpress.com

ISBN
978-1-5255-4964-9 (Hardcover)
978-1-5255-4965-6 (Paperback)
978-1-5255-4966-3 (eBook)

1. SOCIAL SCIENCE, FUTURE STUDIES

Distributed to the trade by The Ingram Book Company

TABLE OF CONTENTS

Two great men joined the heavens during the writing of this book,
Jim and Bill Manewich.

As did a lovely soul who was Jan Scherer, Jeane's best friend.

INTRODUCTION

*The day science begins to study non-physical phenomena,
it will make more progress in one decade than in all the
previous centuries of its existence.*

—Nikola Tesla (1856-1943)

THIS BOOK IS AS MUCH ABOUT HUMAN VALUES, ETHICS
and the big picture as it is about new energy technologies. Game-changing
breakthroughs in physics and engineering are only part of the picture, and
the social mindset needs to catch up with the technical breakthroughs. In
Hidden Energy we bring you up to speed on the next leap for humankind
and why a consciousness shift is beginning.

The famous inventor Nikola Tesla (1856-1943) predicted that in our time
humankind will be stirred up with excitement. He said creativity will be
unleashed in the 21st century at a level dwarfing that seen when his inventions
birthed the Electric Age. Tesla expected a better world when "science begins
to study non-physical phenomena."

He wasn't prophesizing the artificial reality of a digital world. Instead,
he meant studying natural non-physical reality, such as a life force that
exists everywhere.

His discoveries and more recent ones result in inventions founded on
knowledge of vibrations. Both a clean energy revolution and healthcare
can be based on the empowering knowledge that everything lives in a sea
of non-material energy. That paradigm, or worldview, is simple, yet huge
in implications.

Tesla wanted to harness machines to what he called the wheelwork of nature and provide non-polluting electrical power to people at little or no cost. In the final decades of his life, however, he could see that 20th century industrialists—invested in copper mines for vast electrical grids, coal, and oil—had other plans. They would allow only incremental changes to their sources of profit, other than including nuclear fission power plants. And their wealth would allow them to influence governments, publishers and educational institutions and increasingly control the narrative that reaches the public about energy topics.

Nevertheless, Tesla prophesied that the 21st century would bring a more enlightened civilization, not merely new technologies. His dream is overdue.

This book does more than offer hope; it shows that Tesla's dream can come true regardless of the challenging times we are in. *Hidden Energy* tells the human-interest stories of some of today's scientists, engineers and others who, like Tesla, hold a vision of a better world. As they progress toward manifesting this vision, their struggles and successes reveal a larger picture. The human family does not need to accept a dystopian future.

As a possible example of what is emerging, a company called Induction Energy recently announced it is manufacturing a machine called the Earth Engine. It is said to generate, from opposing magnetic forces, mechanical energy equivalent to 40 kilowatts of electricity. That would generate enough power for more than a few homes or enough to operate mechanical devices—without using any carbon fuel.[1]

Our technical advisors have not yet investigated the Earth Engine firsthand at the time of this writing. However, they have tested similar inventions that are working as claimed but not yet on the marketplace. Other varieties of revolutionary clean energy systems are quietly advancing toward commercialization.

The emergence of small but powerful fuel-less energy systems means that humankind's power needs will be met, and uranium and other polluting fuels can stay in the ground. When all countries can enjoy ample, low cost, clean energy without carbon guilt and without radioactive waste, nearly everything changes. Including geopolitical power.

---

1 https://ie.energy/earth_engine.html

Wise planning for that unprecedented shift is urgently needed. The most important planning is to ensure that energy abundance will be used responsibly—not greedily and destructively or weaponized, but instead in service to all life.

Young people, everyday men and women, regional planners, thought leaders and workers in fields including energy, employment counselling, life sciences, agriculture, governance, education, economics, geopolitics, social and environmental justice, spirituality and other sectors can help. No matter what your background, if you are reading *Hidden Energy* you are needed in conversations on how to ethically, effectively and fairly deal with new power-dense energy technologies.

Used wisely, the new energy systems can help build peaceful, just and prosperous economies as well as deal with climate issues and clean up Earth's waters.

Our title *Hidden Energy* refers to:

- A universal energy that can be tapped as a clean abundant power source;

- Nature's ways of quietly working with that life force;

- Little-known technological breakthroughs that operate in harmony with nature;

- The synergy of collaboration by innovators internationally and across generations;

- The divine creative spark in each person. Having hope encourages that spark to find some way, however small, to help create a better world.

No technology is a remedy for all problems, not even a power converter that is small, portable and cheap to manufacture and whose non-polluting output is incredibly energy-dense. Even though it doesn't remove all of humankind's challenges, however, such a disruptive (to corporate monopolies) invention can be the most powerful lever for solving environmental dilemmas, prying society away from carbon fuels, and uplifting a civilization. *Hidden Energy* reveals a variety of such tools that are being developed.

None of the inventions creates energy. The laws of physics state that energy can neither be created nor destroyed. However, physics allows the *converting* of energy from one form to another. Some inventions convert energy from a previously unrecognized source, the background energy found throughout the universe. Some prototypes convert heat from the environment into useable electric power. Some innovate by making microscopic ball-lightning activate in water.

Experimenters use an array of names to describe the source of extra input into new energy devices. They agree that so-called empty space is not empty. Some scientists are saying "dark matter" isn't dark and unknown; many are saying we live in a sea of energy.

When the emerging science becomes dinner table conversation, what will the public call that energy?

Free energy is the most controversial name, yet also the simplest to use. It's a catch-all phrase that people recognize. Yet for many, free energy is a "buzz word" phrase. It has different meanings to different people. For critics who don't investigate, free energy is erroneously taken to mean impossible "perpetual motion" and is therefore dismissed.

Science historian Dr. Peter Lindemann has a definition: "In the simplest sense, free energy is any energy that is provided by the natural world."[2]

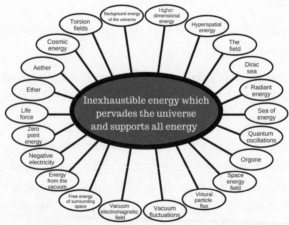

Universal energy's many names. (Image by Jefferey Jaxen)

---

2 Lindemann, Peter, http://free-energy.ws.

We would prefer one simple word or phrase to signify the inexhaustible energy which pervades the universe and supports all matter. However, we respect innovators' choices of words they feel comfortable using, or terms that communicate well with their colleagues. In this book we use each innovator's vocabulary while writing about their ideas. When we choose the terminology, we say universal energy.

In this book, "we" and "us" generally means Jeane Manning and Susan Manewich.

Jeane Manning and Susan Manewich.
(Ken Rochon, TheUmbrellaSyndicate[3] photo)

Susan recently co-founded a foundation that is a community interest company in Europe, to allow for greater international collaboration in the field of new energy technology research and development as well as manufacturing. Earlier, in 2016, the retiring president of the grassroots New

---

3 www.TheUmbrellaSyndicate.com.

Energy Movement had asked Susan to head that not-for-profit educational organization, and its board voted her in as president.[4] She previously had experience with the Resonance Project foundation as a director and member of the Resonance Academy faculty team.

Before those experiences, her professional accomplishments were in conscious leadership development and emotional intelligence. Her clients included Harvard Business School, Yale University, a University of Chicago school of business, London Business School, Singapore Medical School and corporate and non-profit organizations.

Jeane is involved with sustainability projects and new energy innovation through a progressive group of companies, the Avalon Alliance.[5] She has been a social worker, newspaper editor, magazine journalist, and member of the board of directors of a tidal power company. Jeane has written six books about energy breakthroughs and what they mean for society. Her books have been translated by publishers in a dozen languages and sold throughout the world. As does Susan, Jeane most highly values her role as a parent.

Between us, we have a total of more than four decades of exploring an international network of inspired researchers. We have also encountered many entrepreneurs who seek the elusive low-hanging fruit of energy devices that are ready for the marketplace.

Our academic degrees are in social sciences which study how people behave. Topics such as ethics, emotions, motivations and leadership turn out to be relevant on technological frontiers such as the new energy field. Your takeaway from our biographies is that you can approach from your own background, whatever the field, and step into the new energy scene with an intent to help. Society's choice of its energy source is everyone's concern.

An anecdote from Jeane's past further explains why she devoted decades to a field outside her initial profession.

---

4 NewEnergyMovement.org.
5 www.avalonalliance.org.

WHAT MOTIVATES YOU?

The conference hall in Colorado Springs had started to fill. The next speaker was testing the microphone when Jeane heard a brusque question coming from a man who had chosen a chair in the empty row behind her.

"What motivates you to come to these meetings?"

She recognized the man as an engineer from North America who had been a speaker at a conference she attended in Europe. He appeared self-assured and perhaps in his thirties. Now he was leaning forward, unsmiling, waiting for an answer.

To reply with "why do you ask?" would be naïve at such nearly all-male events thirty years ago. A woman who arrived alone couldn't go unnoticed. Many of the male attendees had no more technical background than she had, yet they came to energy-related symposia out of curiosity or for business reasons and could blend in, unlike a lone woman.

"As a journalist I want to educate myself about this field," she replied, "so when the time is right, I can write about it and tell the world some good news."

Her truth was perhaps too simple; he walked away without another word. Jeane appreciated his directness, but he clearly wasn't interested in hearing more. She didn't have a chance to add *I'm here because I love my children. What's in their future? Oil wars? Nuclear waste? Look, I'm a writer. If clean, abundant, cheap, local electricity is possible, I want to tell people. Bring hope. Stop what's ripping the web of life!*

There was a less pressing motive for being there—fascination with the international intrigues involving larger than life characters in the new energy scene. She had briefly considered writing fiction: mad scientist vs. "vulture capitalists" (those who lure inventors into disastrous business entanglements.) Meanwhile, opponents of a new energy source produce mysterious obstacles and threats. Witnessing such machinations can be traumatic, but also provides adventures for a would-be novelist to portray.

Probing more deeply into her motivations, Jeane recalled being elevated to feelings of awe by glimpses of a holistic science. She appreciated the emerging knowledge which recognizes that all living beings are interconnected in a sea of vibrations. She had met frontier scientists who recognize a life force connecting everything in the universe, much like The Force in Star Wars.

Deeper yet was her hope that, as more and more individuals understand the interconnection, many might take a step forward spiritually. There might be more compassion in the world and more of a sense of responsibility to future generations.

No, replying with her deepest motivation wouldn't have been appropriate, decades ago in that Colorado Springs auditorium.

SUSAN'S MOTIVATIONS

When we gave a talk at an Energy Science and Technology Conference in northern Idaho, Susan shared her own motivations, especially hopes for the future wellbeing of her teenage daughter. The audience was predominantly men, yet they applauded the call for more yin energy to balance the "wild West" of the new energy scene. (In Chinese philosophy, yin and yang are contrasting forces that together form everything that exists. Yin is the nurturing principle found in varying degrees in both men and women.)

Susan believes that her path on Earth is to "participate in humanity's transition towards expanding mindfulness." At one level, mindfulness is the practice of maintaining a nonjudgmental state of heightened or complete awareness of one's thoughts, emotions, or experiences on a moment-to-moment basis. In other words, being awake, paying attention, and accountable. Beyond dictionary definitions, mindfulness can be a gentle tuning into other levels of awareness while being fully present with what is happening.

From childhood on, Susan has been aware of how subtle energies influence matter. As an adult, this awareness broadened into seeing the critical importance of how those energies weave themselves throughout the inner workings of human relationships. It became clear to her that the all-pervasive subtle energy, if realized in its beauty and abundance, "has the simple and elegant means to transform and uplift the path of humanity."

The currently dominant worldview on what is possible for humankind is limited and constrained, in contrast to what Tesla and his modern colleagues have envisioned.

Susan found that current research into our sun's effect on Earth signals an upcoming shift at some unknown time. We speculate that the realization about a coming time of great changes could provide the impetus for humankind to align with what is indeed achievable. The way to do that may be to embrace the emerging knowledge and work harmoniously with those subtle yet interconnected energies that you will read about in *Hidden Energy*.

THE FLOW OF THIS BOOK

Part I of *Hidden Energy* touches on challenges faced by scientists in birthing a new paradigm. Part I also introduces what a new worldview could mean in our everyday lives. The story of naturalist and inventor Viktor Schauberger reveals possibilities for working in harmony.

We also follow the travels of Susan and her colleague who is an engineer with advanced expertise in testing energy inventions. He tests devices whose excess output of power cannot be explained by conventional textbooks. The two meet inventors who are private and desire their identity to be kept confidential, but in subsequent chapters you meet other inventors who do speak out openly.

Part II begins with Nikola Tesla, who he was and how his legacy continues today. A Tesla researcher's odyssey leads into the saga of an engineer from India. As had Tesla, he gained understanding about the universal energy both from scientific experiments and from teachers who had studied ancient Vedic texts. The remainder of Part II presents scientists' theories about the abundant energy gift from the universe. A peacemaking engineer has suggestions for defusing what he calls paradigm wars among scientists.

Part III gives a further sampling of the variety of inventions that could make polluting technologies obsolete. Prototypes of new energy devices variously tap into, or are based on, heat from their surroundings, magnetism, aether, gravity, a form of the hydrogen atom or other sources. Part III also touches on the Electric Universe theory which reveals that space is not empty. Part III concludes with an inventor who learned how to run engines on water.

In Part IV we see how holistic concepts play out in the lives of inventors, scientists and other researchers of hidden energy. In the first story, you meet an inventor skilled and respected in electrical engineering who at the same time delves deeply into what he considers most important, the life force. Consciousness and what understanding it means for our future survival is a theme throughout Part IV. The section ends with a highly credentialed physicist's call for broadening of the language and machinery of modern physics in order to come to terms adequately with living systems and life itself.

Part V is about the power of people to create a better world, not only those who solve technical challenges but also those who find solutions to human problems. Part V concludes with suggestions for what you can do to make a difference.

We hope that after reading *Hidden Energy* you will view this era of great changes in a new light. Upheavals can open windows of opportunity. People may exit the cult of money. There will be chances to create a real civilization—steered not robotically but instead by the finest expressions of the human spirit. Let's prepare!

PART I
PARADIGMS

CHAPTER 1
TO FREE THE ENERGY

The New Energy era must ride atop a tidal wave of wisdom.
The stakes are too high for anything less.

—Joel Garbon, industrial scientist,
former president of New Energy Movement

IN *HIDDEN ENERGY* YOU MEET SEASONED PROFESSIONALS, academics and other insiders in the new energy scene. To introduce it we looked for a fresh perspective, and an aerospace engineer and publisher[6] connected us with a student.

Karsten van Asdonk was studying physics while finishing undergraduate courses in biomedical engineering in 2017 when he gave a TEDx style talk at his university at Eindhoven in the Netherlands. Like thousands of other students in his technical university, his appearance could be described as clean-cut casual. He wore a headset microphone and behind him loomed a projection screen. His ten-minute speech was part of a sustainable energy forum at his university.

Karsten van Asdonk

---

6 Dr.Coen Vermeeren,

Van Asdonk spoke in a matter-of-fact and respectful manner. His topic, however, was a taboo. Professors only mention it behind closed doors, if at all.

He introduced himself as "kind of like most of you," an idealist who envisions a future in which all energy generation is 100 per cent non-polluting. "Being this idealistic led me to research our efforts to get there so far. And quite frankly left me shocked with the current state of affairs! Because it turns out that…these beautiful technologies like wind turbines, hydroelectric, photovoltaic…only make up seven per cent of the grand total in power generation."

He pointed to "…a terrible pie chart I made in Paint." A background ripple of laughter indicated the audience was with him.

And the dominant source of pollution, the oil industry, is not losing market share. Instead it is finding new ways to extract oil and to use it, he said. He was left with a big question: "Why is the energy transition happening so slowly?"

"In this research I found out that many inventors over the last century actually have come up with ideas that can outperform most of the technologies we have today!"

He showed a picture of Nikola Tesla, who made progress toward harvesting energy from the vacuum of space. Van Asdonk quoted Tesla, "Electric power is everywhere in unlimited quantities and can power the world's machinery without the need for coal, oil or gas."

Dr. Peter Lindemann explains principles of new inventions (Jeane Manning photo)

After mentioning other inventors who, like Tesla, died without being able to give their most radical breakthroughs to the world, van Asdonk asked "Why are we not driving water-fueled cars? Why are we not receiving electricity from the vacuum of space?"

For answers, he looked to American researcher Dr. Peter Lindemann, who has an honorary Doctor of Science in complementary medicine but is best known as an authority on what Tesla called Radiant Energy. The universal energy that Tesla talked about may turn out to manifest as a life force or an invisible field found everywhere. Today the phrase zero-point energy is more often used. Dr. Lindemann had written an essay about four forces hindering the introduction into the marketplace of breakthrough energy inventions.[7] The first force is the centralized banking system which profits the world's wealthiest families.

Karsten van Asdonk anticipated an audience reaction. "I know what some of you are thinking: 'here comes that guy with…conspiracy theory.' No, we are going to leave that behind."

He instead pointed to what could happen if people built new types of power projects that didn't require borrowing from a bank. A financial system based on the value of petrodollars and benefiting from utility companies' debts would try to stop, or at least regulate, anything that weakens its control. Those debts require interest to be paid to Wall Street banks for creating money to build energy megaprojects.

The second force delaying the public availability of breakthrough energy inventions is national governments. They know that the balance of geopolitical power would shift, or their fuel taxes would have to be replaced by another source of revenue.

One governmental tactic is to issue a restriction on any patent or invention that could give an opponent an advantage. Some inventors in the new energy field have received notice that their invention is under a secrecy order and they can no longer work on the energy device or talk openly about it. They are told it is a matter of "national security." The Federation of American Scientists has compiled statistics on that tactic.[8]

The third group in Dr. Lindemann's essay consists of at least two sectors. One is what he calls 'deluded inventors' who fill the internet with videos of devices that don't work. "The first two forces use the worst cases of these people to promote the idea that any technology concerning this 'free energy' is a hoax, is a scam," van Asdonk explained, "so that nobody believes in it."

"Then there's the people who actually have built a device that works but they're still in it for the money." The student said that sector is vulnerable to being silenced by a vested interest offering a million dollars and the parting message "go live a happy life but don't speak about your invention again."

---

7 http://free-energy.ws/pdf/world_of_free_energy.pdf.
8 https://fas.org.

Van Asdonk asked his audience to guess the fourth force that delays the energy revolution. After a few seconds of silence, someone called out "consumers!"

"Exactly. It's all of us, you and me." Motives of the first three forces are selfish and narrow, but the rest of us share those motives, van Asdonk said. Greed, fear of competition and a lust for power are a common problem.

Regardless, the major change must happen within our lifetime, he told fellow students. "Do your own research into the subject if you are interested. And spread the word. Because it is up to us to develop this technology that is already there. And the first three forces are not going to help us with this."

"My hope for the future is a world in which energy is free for everybody."

His focus is well placed. Nearly a billion people on Earth have no access to electricity at all.[9]

FOR PEOPLE WHO SEEK A BETTER WAY

In a recent conversation, Karsten van Asdonk told us, "I know there's something to do but I don't yet know how. It's not as easy as putting a device out there and making money. The root of the problem goes far deeper, in the psyche of the human, in which there needs to be the change first."

Idealistic youths of previous generations also wanted to change their world. Van Asdonk's generation uniquely feels more urgency, seeing climate change as a crisis situation. He said it is unfortunate that many young people feel overwhelmed by the magnitude of global problems and erroneously conclude there's nothing they can do about it.

Test engineer John Cliss encounters youths who live in despair and poverty, yet they come up with non-polluting energy technology breakthroughs. He had previously worked in the UK defense industry for twelve years with "the best minds that university science can churn out."

---

9 According to The International Energy Agency's *World Energy Outlook 2018* the number of people in the world without electricity fell below a billion in 2017. However, progress continues to be uneven. In 2018 three-quarters of the people who gained access to electricity since 2011 were concentrated in Asia.

> "We are relying on renegade teenagers from the slums of Indonesia and war-torn Eastern Europe for the future technologies..."

"Physicists, engineers, chemists... None of them, as a collective and with virtually unlimited funding, can imagine a technology that some kid in Indonesia with a tin roof can create. It's a marvel of the so-called modern age."[10]

He first experienced culture shock when visiting a small Moroccan village that lacked conveniences most westerners take for granted, such as running water. The children spent their days walking up and down hills to carry water back to the village, yet he knew they had as much creative potential as any child. More recently, he tried to organize travel documents for an inventor whose country is at war. The inventor couldn't fly from his own homeland because the airports are closed. Bombs destroyed many roads, but the inventor somehow had to travel to a neighboring country to get a flight.

Cliss describes the situation as extraordinary. "With all the great education systems in the West, we are relying on renegade teenagers from the slums of Indonesia and war-torn Eastern Europe for the future technologies that the world will run on."

The struggles of innovators are complicated, and implications of an energy revolution are profound, but some basics are simple. Here are brief answers to six single-word questions journalists are trained to ask:

Who? Innovators from a wide variety of backgrounds are making breakthroughs. *Hidden Energy* features inventors who are free from military funding or institutional funding that is connected to the oil industry. Independent inventors struggle to support their own work financially yet are more likely to have freedom to share their discoveries with a grassroots network.

What? A 'free energy' device amplifies or recycles electricity put in to start its processes. In an ideal situation, part of the multiplied power can be fed back in to make the device self-running, Cliss explains.

Revolutionary energy-converting inventions do exist. Some tap into a previously unrecognized source of non-polluting power that is naturally abundant. Some are super-efficient motors/generators, or they recycle electricity. The

---

10   Cliss, John, private correspondence to Susan Manewich, March of 2018.

devices are small in scale and often portable and would be relatively cheap to manufacture, the opposite of energy megaprojects. When developed to the reliability stage, a device the size of a small refrigerator could power a home or neighborhood.

Where? Innovators are at work in many countries, in private laboratories and garage workshops, in a back room of an industrial building, or in a corner of a university laboratory. Energy converters are also rumored to be secretly developed in military-industrial settings, but we have no access to data about those.

When? For more than a century, lone innovators have insisted that they could pull free and useable energy out of their surroundings. Those claims increased in recent years because of information-sharing and advanced electronics and materials.

Why? The inventors and scientists desire what everyone longs for—a chance to create and be successful. They prefer safety, health, freedom and basic comforts for themselves and their loved ones, plus a peaceful society and a regenerated world.

How? The devices take advantage of operating in an *open system* rather than in a closed system.

A closed system can be pictured as being in a box that contains only the machine and its fuel, and the amount of fuel going into the machine is known and finite. The laws of physics that developed in the era of the steam engine deal with closed systems.

On the other hand, the source of power to an open system is not limited to what is known to be in the box. An open system allows energy from the natural surroundings to enter.

Solar panels and windmills are familiar examples of open systems. *Hidden Energy* introduces you to inventions that tap energy from the environment regardless of whether sunshine or wind are available. The breakthrough inventions are small in size and don't require covering acres of land in order to put out enough power to replace fossil fuels. The breakthroughs are in a "small is beautiful" classification.

Revolutionary outside-the-box energy research rarely gets funded, as Chapter Two explains. Regardless, the exploratory spirit burns bright, and experimenters help each other discover basic principles of tapping universal energy.

For the field of research into new energy converters, the online world is both an opportunity and a morass. It is flooded with videos from experimenters of all ages, and the flood deposits some gems. Many of the experimenters, however, have more enthusiasm than patience and willingness to learn and gain expertise. They harm the new energy field's reputation when they solicit donations or investors before having their ideas properly tested.

CITIZEN SCIENTISTS

Certain other open-minded hobbyists who facetiously call themselves shade tree researchers do serious work. Many have spent decades in technical training and relentlessly continue to learn from experimenting. They search the peer-reviewed physics literature for clues on topics that sound exotic to the rest of us, such as "nonlinear oscillations in cold plasma."

They may have day jobs in industry, but their spare time activity does not depend on government or industry grants. Therefore, their research doesn't have to focus on incremental improvements to existing technology and doesn't have to avoid disrupting corporate interests. Instead, the hobbyists are free to focus on bringing clean power abundance to the people. Unless they run out of money for materials, machining or instruments.

These inventors use or combine actions such as:

- Employing nature's tool, the vortex, in converting energy.
- Producing powerful effects like a miniature tornado would.
- Understanding electrostatic, magneto-electric and other effects.
- Capturing & reusing spikes of electricity for a power gain.
- Timing pulsed electricity for big output from tiny input.
- Recycling the electricity within a motor or generator.
- Separating functions of motor/generators to evade counter-force.

- Using super-fast electronic switching.

- Designing with special geometries and natural ratios.

- Drawing universal energy into a system via toroidal motion.

- Coupling non-linear devices into a series resonant circuit.[11]

- Discharging voltage abruptly into an unsettled plasma.

- Imploding microscopic bubbles in water (cavitation).[12]

- Making 'motional magnetic fields' self-oscillate in magnets.

- Working with gravity or 'electro-gravitics.'

- Creating excess heat via unique processes neither chemical nor nuclear.[13]

- Bringing in experts in nano-technology and materials.

- Innovating with advanced alloys such as Nitinol (nickel-titanium).[14]

- Learning "there's no direct relationship between the strength of a magnetic field and the quantity of electricity used to produce it."[15]

Varied breakthrough energy inventions exist in many categories; it's not a competitive case of one technology being the winner or one person single-handedly saving the world. The new energy field can be compared to a forest with diversity—pine, birch, cedar, willow, larch and other species that support each other. Trees in a natural forest connect through their roots and miles of tiny filaments and are not just a monoculture as in a tree plantation. Each variety has its place in the sun or shade in an interconnected ecosystem.[16]

---

11   The Magnetic Resonance Amplifier invented by Joel McClain and Norman Wootan.
12   https://nanospire.com.
13   www.lenr.org or coldfusionnow.org.
14   http://electricityfromair.com.
15   Paul Babcock, inventor, see chapter 21 for his findings.
16   Forests provide an analogy, exemplifying diversity. Professor Hans Pretzsch, co-author of an international study in which the Technical University of Munich participated, says mixed-species forests are more valuable as versatile habitats, mitigate climate change and are more stable against pests. They're better supplied with light, water, and soil nutrients via their complementary crown and root systems, which makes them resilient during dry years. Diversity has advantages.

American inventor John Bedini experimented
with novel energy devices (Jeane Manning photo)

Mindful planning means keeping in mind that diversity furthers long-term success, whether in life forms, technology, an economy or society.

STUDENTS ARE OPEN TO NEW PARADIGM

After university student Karsten van Asdonk spoke about breakthrough inventions at the sustainable energy forum, most students' comments were positive. On the other hand, professors were silent, except for one who identified as a faculty member but didn't sign a name on the note. That anonymous professor wrote that Karsten van Asdonk should not have presented at the university because his topic was "pseudoscience."

"I expected that," van Asdonk told us. "The university is like an impenetrable fortress, a hierarchy. You have to obey some kind of structure and cannot just research anything you want."

The main paradigm in universities is still a materialistic one, van Asdonk believes. "To admit that there are phenomena which we still cannot describe even with quantum mechanics or string theory is a whole different step. Because scientists like to calculate things. They want to have models which predict the outcome of an experiment."

"When an experiment becomes subjective, for example a spiritual experience, you cannot hang a meter on that. You cannot measure what the experience is like."

The scientist you will meet in Chapter 2 does expand the boundaries of knowledge. He is a distinguished academic leading the way to a new paradigm about water. The challenges he's encountered are similar to obstacles met by people involved with revolutionary energy devices.

CHAPTER 2
REVOLUTIONARY MEETS
GATEKEEPERS

What's at stake is the very future of the world, for a world
without scientific leaps is a stagnant world,
susceptible to decline.

—Institute for Venture Science[17]

CREATIVE STUDENTS AND OPEN-MINDED SCIENTISTS ARE
likely to thrive in the Pollack laboratory at the University of Washington in
Seattle. "If you believe that everything in the textbook is correct, then you
would probably do better elsewhere," says the lab's website. On the other
hand, those open to discovering new fundamentals of science may find the
laboratory to be a compatible place.

Gerald H. Pollack, PhD, is the bioengineering professor in charge of the
laboratory. He is a tall gray-haired scientist whose expression often emanates
kindly amusement. Despite honors and accolades received, he seems to take
himself lightly and find the humor in life.

His book *The Fourth Phase of Water: Beyond Solid, Liquid and Vapor* reveals
discoveries about a life-enhancing phase of water he calls EZ (exclusion zone)
water. It has implications ranging from biology to energy and information

---

17   https://ivscience.org.

storage, and even getting electricity from water.[18] The essence of EZ is separation of electrical charge, he explained in a 2018 Energy Science and Technology Conference panel discussion.

"EZ has negative charge and you've got positive charge next to it. When you have huge amounts of negative and positive charge sitting next to each other you have huge amounts of potential energy."

One of his favorite quotes is from the father of modern biochemistry, Albert Szent-Györgyi, who said "Life is water dancing to the tune of solids." Dr. Pollack's own words also resonate. He concluded a talk at a large university with, "What we like the most is understanding the gentle beauty of nature."

His book about water is internationally making waves, forgive the pun, as are the water science conferences he started. Being a pioneer, he encountered opposition that is mirrored in other fields including energy.[19]

You don't expect gatekeepers in science, passing scornful judgement on something they have not investigated. The scientific method is all about investigating. It includes precise, accurate experiments and reporting all observations, especially anomalies. Exploring the unexplained has in the past led to new worldviews, but today it's often hampered.

DR. POLLACK HAS A BOLD SOLUTION.

From his home base in Seattle, Dr. Pollack founded a non-profit Institute for Venture Science to support breakthroughs—ranging from health discoveries to new understandings about electromagnetism and gravity. The institute's job is to nurture revolutions in basic science, not fund technology.

The professor didn't start out as a revolutionary. He describes what he was like as an undergraduate electrical engineering student—properly dressed and duly respectable, wearing the conventional suit and tie like his peers.[20] Nevertheless, seeds of rebellion were planted.

---

18   Pollack, Gerald H., *The Fourth Phase of Water: Beyond Solid, Liquid and Vapor*, Ebner and Sons Publishers, Seattle, 2013.
19   Ibid., p. 339.
20   Pollack, G.H., preface to *The Fourth Phase of Water*.

Early in his career he studied how muscles contract and found evidence that accepted science about it is incorrect. Then he watched established scientists in that field cling to their theory even after it was disproven, as if adhering to a religion. Eventually he noticed that in every field in science, evidence against the prevailing view is often ignored.

"It's easier to go with the mainstream views and get accolades from the leaders in a field than to challenge those views. Because those leaders have so much power…That's not how science should work."[21]

Resisting breakthroughs is natural, the Institute for Venture Science website explains. For instance, imagine leaders in physics being confronted with compelling evidence for a form of energy previously unknown. It may be exciting for the physicists, but also feel destabilizing and threatening to their self-interest.

"As a result, temptation exists to dismiss the challenger with a mere wave of the hand. We may even hear 'pseudo-science' or 'crackpot science.' Such casual dismissal lends reassurance to members of the group under challenge; they can feel secure in their thinking."

In the past, challengers of a scientific paradigm only needed to sway a small number of influential leaders. Today, success requires convincing large numbers who may be reluctant to change their views. Gatekeepers control public opinion and funding.

Yet, hasn't science progressed substantially in this century?

Dr. Pollack points out that yes, novel technologies arrive, but new *basic knowledge* has become rare. For example, the war on cancer was declared more than 40 years ago. At that time, a woman with breast cancer had only a few choices—surgery, radiation, and chemotherapy. Despite the billions of dollars poured into the war on cancer, today's therapies are the same choices with some improvements.

Breakthroughs as game-changing as splitting the atom, penicillin, or a polio vaccine are not showing up.[22] The laser, the transistor and the Internet were based on revolutionary science—from more than 30 years ago.

---

21   Pollack, Gerald, interviewed by Jeane Manning, Nov. 28, 2018.
22   Pollack, Gerald H. Letter to US President Barack Obama, July 14, 2009.

After over a century of polluting, the energy industry is still consuming dirty fuels including uranium. Windmills and solar power are improved but not new. Tidal power is barely tapped.

Press releases about findings that promise to cure cancer or revolution-ize energy often comes from institutions that gain prestige by putting out an exciting announcement. A university's publicity department can give the impression that their science laboratory discovered a world-changer. However, professor Pollack explained, often the laboratory's scientists are only suggesting a possible application for a finding but have not yet had that particular success.

> **Breakthroughs were made by scientists who were free to explore...**

In a letter to former president Barack Obama during Obama's administra-tion, Dr. Pollack bluntly described roadblocks to science breakthroughs. The letter was also signed by other distinguished scientists. It explained that, in past eras, breakthroughs were made by scientists who were free to explore wherever their curiosity and unexpected observations led.

Today, administrators decide which research areas are likely to be fruitful, then the researchers must show progress within their niche or risk losing their paycheck. They usually make incremental progress, which means moving forward on the same path in small steps rather than stepping aside to investigate an unknown that could overturn our thinking about some-thing important.

The process of vetting (deciding whose work deserves funding) is weighted. When someone applies for a grant, its administrators appoint an estab-lished leader in that field to review applications. Such leaders are naturally inclined to support applicants whose views agree with their own. Grant applicants understand this, Dr. Pollack explained, so they submit safe incre-mental proposals.

The scientists' letter to the White House said a solution requires dealing with those "elephants in the room: the mainstream scientific leaders who are reluctant to consider ideas that threaten their own well-funded efforts, which invariably represent the status quo." The letter proposed an independent institute, unencumbered by past philosophies and procedures.

No one in government released money for it. The Institute for Venture Science began anyway, relying on private donations. It has received enough to start reviewing proposals.

The institute has a strategy for a second step. It involves critical mass— funding at least ten scientists or teams, in separate laboratories worldwide, who have expertise for a given line of research. Each might use a different method to check the work on a given hypothesis and they would come to independent conclusions. (A hypothesis is a tentative explanation for an observation, phenomenon, or scientific problem that can be tested.)

Depending on their findings, the scientists will either abandon the hypothesis or take it seriously. It would be difficult for the science establishment to avoid talking about a disruptive discovery if a dozen research groups with complementary expertise had validated the same potentially earth-shaking idea.

"Until that happens," Dr. Pollack said, "even the most compelling of revolutionary ideas will languish in obscurity, as many now do." Lone dissidents have been ignored by way of dismissals such as "everyone knows Dr. So-and-so is deluded; that's impossible, don't pay attention."

The professor who dances with water doesn't aim to foment scientific revolutions for the sake of being disruptive. The goal of the institute is to restore science "to the richly bountiful enterprise it was before the funding agencies began imposing top-down management and inviting mainstream scientists to judge their challengers."

WHAT ABOUT NEW ENERGY GENERATORS?

One of the experts endorsing the Institute for Venture Science noted that the science community still does not understand where gravity and electromagnetism come from and said the institute could break through such conceptual barriers.[23] Dr. Pollack explains why that is important. "Technologies of today

---

23   William A. Gardner, originator of the Cyclostationarity Paradigm, https://ivscience.org

were based on the fundamental discoveries of yesterday. So, if we want to keep going with technologies, we need the underlying science to provide the platforms..."

Electronics pioneer Ken Shoulders (Chapter 19) straddled both basic science and technology in his work. He once commented on problems faced by explorers of "free energy machines, 'cold fusion' or nuclear waste remediation." Attaching the explorers to institutions was not the answer, he said, because it could take away their freedom to roam into uncharted and unexplored areas.[24]

---

24   Shoulders, Ken, 2004.

CHAPTER 3
HARMONY WITH NATURE

How could we have missed this universal machine; why have we ignored the vortex, the workhorse of the universe?

—William Baumgartner, mechanical engineer.

VIKTOR SCHAUBERGER WAS ON SUSAN'S MIND ONE DAY while hiking on the Hawaiian island of Kaua'i. She and two friends were approaching the heart of the island at Mount Waialeale waterfall. That remote area, and the seashore where she surfed, both exude a feeling of vitality. It's an almost tangible life force pulsing in the misty air, in leaves of the trees, and through the mud under her bare feet.

The Austrian forester Schauberger (1885-1958) would have understood. He studied how natural energy moves in the air and water and was fascinated by the energetic potential of cold pure water in mountain streams, for instance. He developed science and technology based on comprehending nature.

Susan's hiking companions were two women, longtime Kaua'i residents, who shared the reverence for this island. While they ascended narrow trails lined by blossoms and spiral-twisted tree shapes, Susan recalled that three-dimensional spirals seen throughout nature were a key to Schauberger's understanding of energy.

The hikers were far from roads and gasoline fumes. Only birdsong, rivers gushing, and the rustle of leaves punctuated the silence. Susan had read that Schauberger was fiercely protective of wild rivers, streams and vast forests in his mountainous homeland, so he invented technologies to rejuvenate rather than deplete the health of soils, plants, water and atmosphere.

Schauberger's inventions and writings showed that humankind could enjoy technology—for transportation, for comfortable buildings and for abundance of electrical power—*and* be in harmony with nature's subtle energies. However, he had warned that a society would first have to rethink its values.

Susan's attention returned to the heart of the island, its deep peace. She and her hiking companions saw the island as a microcosm of the world; it holds tensions and potentials found in human society around the planet. On Kaua'i, wealth in the form of expensive lifestyle concentrates on the North Shore. Susan had noticed that much of the moneyed sector on the island ignores a rich heritage nearby, the culture that respects nature's creativity as sacred. At the homestead where she had been staying, her hostess and host have deep roots in the land. Their Indigenous culture, she reflected, is grounded and refuses to be swept away by the flood of materialism.

On the west coast of the island, industrial pollution affects the health of already-marginalized Indigenous people. Upwind from their homes, four out of the five major international biotech companies—BASF, Dow, Pioneer, and Syngenta—managed 15,000 acres. On that land the corporations stored pesticides and experimentally crop-dusted the toxins onto plants. The state of Hawaii and the federal government welcomed those multinational corporations, but many local residents—and whistleblowers with science credentials—saw differently.[25]

Schauberger had pointed out that people whose industries and endeavors are in harmony with nature are more likely to be in harmony with each other.

---

25   Film maker and Kaua'i resident Keely Shay Brosnan chronicled those tensions in her documentary *The Poisoning of Paradise*.

WHY HEED A LONG-DEAD FORESTER?

The Viktor Schauberger biography *Living Water*, by a Swedish electrical engineer, changed Jeane's life in 1986.[26] It gave her hope that new energy technologies could help rejuvenate air and water instead of deplete and pollute. And that basic concepts of how humankind could switch to regenerative technologies are simple enough for a non-technical person to easily grasp.

Schauberger had studied how hidden energy from the cosmos enters into nature's spiraling motions, as in flowing water, air currents, and other natural vortices. His observations resulted in energy-harnessing discoveries. But most people missed the cue.

How did he know so much?

He came from generations of Austrian foresters. His family's motto translates to "faith in the silent forests," and Viktor became the most faithful and watchful of all. He was exceptionally perceptive, and from childhood onward he wanted to grasp some elusive truth he wasn't finding in school or church. His father sent Viktor's brothers to university to become professionals, but when Viktor turned eighteen, he refused his father's directive. Biographers say he didn't want his perceptions corrupted by people who are alienated from nature. After his brothers attended university, their thinking seemed conformed and compartmentalized.

The rebellious youth left home to live alone in a remote forest. An Austrian prince hired him to look after royal land holdings in a mountainous area without roads, larger than almost any pristine forest that exists today. Unlike standard laboratories for studying science, the forest setting allowed him to closely observe the workings of a vast intact ecosystem, for years.

He became aware of water's special role as the lifeblood of the planet. He watched rare phenomena, such as a landlocked lake rejuvenating itself by suddenly creating a fast-accelerating whirlpool followed by a massive waterspout, a vortex rising above the lake.

A *vortex*—three-dimensional spiral—shape is basic to the movements of water and air, he learned. He also noticed the breathing motion of water in

---

26   Alexandersson, Olof, *Living Water: Viktor Schauberger and the Secrets of Natural Energy.* New edition published by Gill & MacMillan, 2002.

a natural stream—whirling and drawing in air whenever it encountered a protruding rock.

Temperature plays a role in what he called "living water," more than hydraulic engineering experts realized. Schauberger noted the trees and bushes on the banks of mountain streams or rushing rivers; the vegetation shades and cools the water. At nights, in the light of a full moon, he learned about the heightened energy state of cold pristine water by seeing sparks of light in the water and certain egg-shaped rocks floating. Different shapes interact with energies uniquely, he noticed.

One day he startled a large trout in a swiftly flowing stream. He'd been wondering how the fish could remain motionless in fast-moving water with only slight movement of its tailfins to keep its position. How did it flee upstream instead of letting the current help push it downstream?

The observant forester figured out that the fish's shape and motions caused vortices to form and push the trout against the current. Schauberger's experiment—pouring hot water into a creek at a distance upstream from trout—caused them to flounder and be pushed downstream. It showed a relationship between water's temperature and its ability to form energetic vortices. But he had yet to learn another key concept.

On a cool moonlit night one spring, as he sat beside a waterfall, he noticed a large fish darting back and forth in the river in twisting motions as if building up energy. Suddenly it disappeared up into a waterfall's huge jet of falling water. He caught a fleeting glimpse of the fish spinning wildly under a cone of water and then floating upward until it tumbled over a curve at the top of the waterfall. As a result of certain lighting conditions that revealed the path of what he later called levitation currents, Schauberger had seen the fish rise as if through a tube within the misty veil of the waterfall.

He concluded that even while gravity's pull on water creates a visible downstream flow, invisible currents are going in the opposite direction, upstream, in a river in its natural state.

To help people imagine such a levitation force, in his writings Schauberger suggested first visualizing a whirlpool. Like water swirling down a drain, a tunnel forms in the middle of that vortex and sucks objects downward. Then imagine a *whirlpool turned upside down*. Instead of being pulled down, a trout in an upside-down vortex would appear to be floating upward along the axis of the vortexian spin.

QUIT JOB, STARTED INVENTING

Schauberger quit his job as royal forester when his employers began to log the forest greedily instead of selectively. By then he understood principles he used later in his "biotechnical machines" which bore no resemblance to invasive technologies promoted today. Schauberger's inventions ranged from a copper plow for agriculture that caused no harm to the skin of the earth, to an implosion-based generator for powering a house.

He invented unorthodox water pipes to protect and enhance energy that's built up within layers of naturally flowing water. The pipes had an egg-shaped cross section and were twisted along their length. The twists caused water to spiral inwardly toward the center and pull away from the pipe walls so water could move faster and with less friction. Water was even *sucked forward* with increasing speed and no longer needed to be pushed by gravity or pumps. It was 'vacuum power' instead of pressure.

Inward-spiraling (centripetal force) as well as centrifugal force became a key concept in his energy generators. He said inward-spiraling motion enhanced the natural life-force in water or air; it concentrated the substance instead of expanding outwardly to dissipate it. Schauberger pointed the way toward energy systems that use nature's quiet, cooling implosion forces.

COPYING NATURE'S MOVES

His inventions were unusual. Other industrial engines such as the one under the hood of a car rely on explosion. Explosion releases heat to cause expansion and pressure for producing work such as moving a piston. Schauberger saw that nature uses the opposite process, cooling, to cause suction and vacuum forces. A tornado is the extreme, and destructive from our viewpoint on the ground, example of nature's use of suction to produce work.

Schauberger was first to build working models of what are called implosion turbine engines. Some of his inventions created vortex motions in water, then a combination of gravity and centrifugal force produced a continuous motion in those systems without burning fuel.

Implosion-based systems tend to increase order instead of disorder. Nature uses the inward-spiraling motion to quietly create, but also uses a noisy or heat-producing and expanding type of force wherever it needs to break things down or fling a substance outward. Decaying, composting, burning, dissipating or destroying material is only one half of nature's eternal cycles of death and rebirth.

Schauberger warned against splitting the atom for nuclear power plants or burning coal, oil or gas to power society. He predicted destruction from too much explosion-based technology—unbalanced, the natural world would degenerate, and people might morally degenerate along with it.

He urged people to comprehend nature and copy nature in their technologies. His grandson Jorg Schauberger adds to Viktor's adage by suggesting that people comprehend nature, copy it and then *cooperate* with nature.

RIGHTS OF NATURE

On Kaua'i a woman put a conservation easement on land that's registered in her name, to protect nature on that land perpetually. Although few people are able to do as she did, many recognize that ecosystems—trees, oceans, animals, mountains—have rights to exist and maintain and regenerate their vital cycles just as humans do.

Viktor Schauberger would applaud.

His understanding of earthy topics such as trace minerals, diamagnetism, magnetism and life force in the soils will be priceless in the near future as climate change affects agricultural land. Likewise, in a world of increasing competition for drinking water, Schauberger's knowledge of how to restore the health of rivers and of the hydrological cycle is vitally needed. He deeply studied water's journeys into the atmosphere and back to the ground and up through the trees again into the atmosphere.

The wealth of information in a series of books by Schauberger's biographer Callum Coats could be the basis for an eco-technology degree program at a new paradigm university.[27] Another book, *Hidden Nature,* by one of Coats' publishers, summarizes the concepts in contemporary language.[28] Schauberger's insights are guideposts to a mindful path toward a society in harmony with the natural world.

---

27   Coats, Callum, *Living Energies*, Gateway Books, Bath, UK, 1996 and other books by Callum Coats.

28   Bartholomew, Alick, *Hidden Nature: The Startling Insights of Viktor Schauberger*, Floris Books, UK, 2003.

CHAPTER 4
A SEARCH FOR THE
ETHICAL WAY

*In the future machines will be driven not only by water and
steam, but by spiritual force, by spiritual morality.*

—Rudolf Steiner [29]

JOHN CLISS IS A PROFESSIONAL ENGINEER WHOSE CAREER
started with twelve years as an R&D (research and development) scientist
in the UK defense industry. He worked on highly classified programs, those
officially designated Above Top Secret.

During his time in the defense industry he
had an intense personal transformation; the
ethical epiphany concerned his job which
involved weaponization of advanced technologies. He decided to leave that industry.

He considered building a new career
around renewable energy systems. Improved
solar and wind power were finding new
markets. However, he realized they won't
quickly enough make the huge impact
needed to stop oil wars and pollution. The
fossil fuel industry looms large in the world;

John Cliss, engineer
and scientist, UK

---

29   Steiner, Rudolf, January 2, 1906 Berlin.

renewables are like mice who think they can overthrow an elephant by reproducing prolifically.

Instead, John Cliss saw that his professional skills could help in the search for what are popularly called free energy devices or Tesla technology (not to be confused with the Tesla electric car company.) His research between 2009 and 2014 revealed a need for professional standards for evaluating and documenting whether such devices work as claimed. The only way to be certain whether they were real was to get his own hands and test instruments on them.

In 2014 when he left his secure job, Internet forums were buzzing about an open source energy device to be built by the people, for the people. Online interviewees claimed a QEG (Quantum Electrical Generator) created ten times more power out than power in.

The first QEG was to be assembled in a small Moroccan village in April that year; engineers had been invited. The open source aspect of the project appealed to him. He knew that any technology with a radical implication for the defense industry could be classified immediately.[30] If the inventors had applied for a patent instead of openly giving out information, they would have received a secrecy order forbidding its manufacture.

On the spur of the moment he booked a flight to Morocco and packed his $100,000 worth of electronic measurement equipment—oscilloscopes, voltage probes, current probes, temperature guns, spectrum analysers—into a large suitcase. Spending his own money for travel, he wanted to be certain whether the device worked as advertised.

In the Moroccan village about sixty people from nearly thirty nationalities were ready to join the project. His impression was of a group of wholesome and good-natured people—engineers, spiritual gurus, New Agers, hippies, and dropouts from the conventional business world wanting to be part of something new. They were refreshingly different from the conventional engineers he had worked with for twelve years.

After hardware parts arrived, over a few days the engineers assembled the machine. Cliss particularly questioned how the reported over-unity (more power output than the operator put into the machine) had been measured

---

30   The authors use American spelling, defense. John Cliss, from the UK spells the word as defence.

on the original device in the USA. He felt uneasy about some of the replies. Meanwhile, the QEG drew attention across the internet, with crowd-funding for air tickets and parts.

Intense expectation developed, both online and in the hot, confined work-shop space as the device was readied. The feeling was similar to waiting for a baby's arrival. People poked their heads around the door, asking "how long?" and "is it finished yet?" Pressure built up. The assemblers were told that even the king of Morocco had given his blessing to the project.

When they finally powered up the device, the motor's rotation made a loud humming, throbbing sound. A flash of lightbulbs indicated power on its output side. The crowd responded jubilantly with kissing, hugs, and tears. A pleasant feeling permeated the workshop, a sense of having collectively achieved something.

John Cliss and two other engineers then worked day and night to fine-tune the device and perform high-quality measurements of power in and power out. The builders hoped that power out would be greater, so it could be called over-unity, free energy, or CoP greater-than-one. (CoP stands for coefficient of performance.)

Cliss had the specialized equipment for such measurements[31] but was concerned about how the ten times more power out than power in, reported earlier by others, could have been assessed without such equipment.

"That was my first clue that all was not well," he told us.

Online announcements that the QEG works and was producing over-unity made him uncomfortable. After about a week, a backlash hit bloggers; others in the online community asked for proof of measurements.

John Cliss' equipment measured the QEG at no more than 90 per cent efficient, meaning that for every unit of power put into the system, only nine tenths of a unit was coming out. It was an efficient transformer, and *not* a free-energy device. Such subtleties seemed lost in the online excitement about the device.

After days of tweaking and tuning the QEG, over-unity was still not forthcoming. Cliss realized what had gone wrong. When the QEG had first

---

31 With the QEG, the frequency of operation of one of the circuits was into the kilo-Hertz (kHz) range, and voltage was more than 10,000V, meaning that specialists' equipment was required.

been presented to the public, it was described as a free-energy invention with a working device in the USA that would be replicated in Morocco. "This was my first hard lesson in miscommunication. 'Working' was never fully defined."

He learned that the claims of tenfold output came from the original builder of the device.[32] The QEG team only had the motor spinning which they called "working" but had never had it at over-unity themselves. The QEG, as far as John Cliss knows as this is written, is only a research and development project. He has no problems with that if false claims are not made.

Cliss has mixed feelings about the project. He resonated with the intent of everyone involved to open source it and help the world. On the other hand, the project was based upon an untruth because the device put out less power than it took to run it.

Even though the QEG did not work as claimed, Cliss said he holds pleasant memories of the experience and the people.

'WILD WEST YEARS'

What he calls his Wild West years were from November 2014 to December 2017. He uses that phrase because he noticed no rules on the frontier of new energy science. Like other test engineers who investigate, he had to pursue dead-end leads when misinformation had been hyped. The unfunded frontier falls outside of academic peer-review structures that are supposed to stop misinformation. However, he did not want to merely criticize "because there was a pioneer spirit about these swashbuckling inventors who want to change the world for the better."

He was able to connect with an investor who had for years been involved with energy projects despite often losing financially. Cliss learned there are investors who want to help but often lack the expertise to decide what technology works and what doesn't.

The next three years he travelled internationally, testing devices. He began by reaching out to inventors who had posted YouTube videos that

---

32   An organisation called WITTS had claimed over-unity.

demonstrated promising work. Visiting them and making power assessments gave him a network, and his professional reputation opened further doors.

Sometimes he was asked to test a publicly known device such as the QEG had been. During those years he usually found the inventor had made an error in measurements.

Disproving dozens of devices was not work he enjoyed but was necessary. Errors are easy to make; most inventors have neither the budget nor the expertise for making high quality measurements. It requires expensive equipment, especially if they built a system involving high frequency or high voltage electricity.

The first few years of Cliss' quest for a convincing device were discouraging.

He caught a glimpse at a small company in southern Europe where an inventor demonstrated such a machine. The inventor and team had done professional quality work building a device that they said put out about three times more power than was input to run it.

Cliss tried to get investment money to them, but the amount they wanted for the technology had been more than the potential investor expected. Cliss' main disappointment was the European inventor refusing to allow him hands-on testing with his own calibrated equipment.

SYNERGY IN BUSINESS PARTNERSHIP

In his travels John Cliss met Susan. When Jeane later asked how, he said he had been invited to Germany for a workshop. Its personal development topic was outside of his comfort zone, but he attended on a whim. As a conventionally trained engineer he had never been taught to "go quiet, tune in, and listen." However, during the four days he was surprised at the information that came through.

He had been standing on the lawn outside the workshop room when his host said, "there's someone I want you to meet." For some reason that Cliss didn't understand, he knew he had to meet the person named Susan. They spoke on a group call and he learned she is a leadership consultant and president of the New Energy Movement.

Days later they talked again, about similarities that would lead them to collaborate. What really connected them, he recalls, was that for eight years he had been studying the work of Austrian philosopher (1861–1925) Rudolf Steiner. Steiner wrote about what he called 'Moral Technology.' Steiner said that in the future humankind would move beyond the electronics era to a type of technology that would not perform unless the "moral consciousness" of the operator was at a high enough level.

At first Cliss couldn't understand that. He had been taught that a scientific experiment is repeatable by another person regardless of anyone's moral rightness. Then he expanded his reading list back to the 19th century American inventor John Keely and could start to imagine such a technology.

John Ernst Worrell Keely (1837-1898) worked with forms of vibration to set machines into motion. To picture Keely's sensitive machines, imagine musical instruments built to respond to tones that are in harmony and also built to note the builder's breathing and brainwave rhythms. It was as if a violin could only be played by the person who made it.

Dr. Dawn Stranges, behind the sphere, helped Dale Pond recreate a Keely invention. (Jeane Manning photo.)

Keely gained some mastery over a force he called the ether, but by then he had run out of money. Acquaintances organized a company for him, but dishonest promoters sold stocks and pocketed the proceeds. Nevertheless, Keely discovered more than forty of what he called fundamental laws of nature, long time researcher Dale Pond tells us.[33] Keely apparently understood the subtle nature of vibration, cycles and rhythms, and of relations between the worlds of matter and of mind.[34]

When Cliss heard about beings who are countless years ahead of Earth's inhabitants and operate their space ships by tuning into them, he realized

---

33   http://www.pondscienceinstitute.on-rev.com.

34   See our chapter-length article on John W. Keely at http://HiddenEnergy.org.

that the information "mapped" with Steiner's work. *This absolutely fits the future,* he had thought. *Free energy is almost nothing compared to what moral technology is. Free energy is only a tiny whisper of a first step. And yet we human beings are still tripping over ourselves trying to get there.*

John Cliss and Susan Manewich each want to help inventors and investors move forward in an ethical way. Their similar ideals resulted in a bond of friendship between them, and a professional relationship.

THE TEAM'S FIRST ADVENTURE

A journey in Latin America was their first joint adventure of meeting with an inventor. Cliss invited Susan to accompany him on a trip in December of 2017 to check out an invention. She boarded an airplane in Boston, and he flew from London, to a smoggy airport in Latin America.[35] It yielded a turning point. To tell that story, in the next chapter we switch to the informality of using first names of both Susan and John.

---

35   A non-disclosure agreement requires keeping the location and the inventor's identity confidential.

CHAPTER 5
HUMBLE START TO
PROFOUND SHIFT

*Most evidence of the possibility of radically different and
better technology is not actually hidden, but simply ignored
and belittled by the mainstream...*

—William Zebuhr, PhD, technical editor
Infinite Energy magazine.[36]

WHEN SUSAN FIRST WENT TO SEE A WORLD-CHANGING
invention, she didn't anticipate that stray animals would give her an insight
about it. The location was a humble neighborhood in a Latin American country.

Their host, whom we will call José because his team protects their privacy,
had chauffeured Susan and John on the five-hour drive to the modest laboratory building. When Jose picked them up, John first had to load thousands of
dollars' worth of his measurement equipment into the car. Customs officials
had charged John a hefty fee for bringing so much technical hardware into
the country, but it was necessary.

José opened up during the car ride. He said the inventor they were going
to see had worked for decades to build a special electronics board, small as
a shoe box, that powers a motor. It works by harnessing a force called back
EMF, the electromotive force that ordinarily opposes the motor's torque or
turning power. The Latin American inventor's early attempts had produced

---

36   Zebuhr, Willliam, Nov./Dec 2016, Issue 130, Infinite Energy magazine, p. 6.

only nano-watts of excess power, but over time he developed the concept into a device that produces kilowatts—enough to power a house, car or small factory. John was anxious to find out if those claims would stand up under rigorous testing by his array of instruments.

José told them about the many potential investors that the inventor and his team had entertained during the past fifteen years. He said some investors "get the Gollum look in their eyes" (a *Lord of the Rings*[37] reference to greed that surfaces when some imagine becoming wealthy in the trillion-dollar energy industry). Suddenly the investors want "all the marbles" instead of dealing with the inventor in a fair and balanced way.

When the car drove through the gate at their destination, a large shaggy dog and a cat with kittens were wandering around a dusty yard between a nondescript wooden building and a chain-link fence topped with barbed wire. José parked on the dry grass inside the gate.

The inventor and his engineer and assistant greeted the visitors. Susan and John were led inside the laboratory building to a small office where the inventor, cigarette hanging from his mouth, explained his machine. He presented output data from different stages of the device's development, up to the latest reading of four times more power out than anyone put in to run it.[38]

While José and his partners ran the device and watched John measure and record data, Susan pondered a different set of anomalies: the inventor continued to light cigarette after cigarette, yet Susan's body didn't react as negatively to smoke as usual. Instead, she had a subtle feeling of well-being despite the smoke-filled dry air and thirst from traveling.

And the loud motor being powered by the inventor's electronic device didn't frighten the animals. Instead, the scruffy dog, tail wagging, and the mother cat with kittens came in the open doorway and wandered toward the machine before it was turned on, as if they perceived the man's intention to start it. They edged closer to the machine. When the inventor switched off his machine, the animals soon left. Susan puzzled over their behavior and remembered that animals have sensitivities such as being attracted to drink fresh water that outwardly looks identical to a bowl of slightly stale water

---

37   *The Lord of the Rings* is J. R. R. Tolkien's epic fantasy novel.
38   CoP is a term used in heat pump and other industries and stands for Coefficient of Performance.

nearby. *Could the device have made some connection with the life energy that felt good?*

Later that day, John confirmed that the device operated at over-unity, meaning more output than anyone had put in. He celebrated being able to "sign off" on his own measurements of such a revolutionary device.

At the time of this writing, the government of the inventor's country is investigating that device as a way for an industry with a heavy demand for electricity to save power. Non-disclosure agreements prevent us from giving further details.

John had little chance to sit back and savor his first "find" of a working device; the next month's travels caused him to exclaim that floodgates are opening in the energy-breakthrough scene.

THE 'BOUNTIFUL YEARS' BEGIN

In January of 2018 John Cliss was called by an investor who had heard of him and needed an independent third-party report on a device in South-East Asia. John was given airline tickets but little information. After several connecting flights he arrived at a hotel in a large third-world city.

Slightly apprehensive, he met with the project's team. Local helpers wheeled a box, about a cubic meter in size, into the hotel's private meeting room. John was not allowed to see inside the "black box;" he was only to measure the power going in and the power coming out of the box.

The device generated kilowatts of output, more than thirty times more power out than the operators put in. As a cautious engineer, John adds to his conclusion the phrase "unless batteries were somehow hidden in the box," although the vetting team closely inspected it visually afterward to ensure no foul play.

"It was an extraordinary moment for me personally," he reports, "and a great relief I suspect, to both the inventor and investors to have a third-party independent validation. In fact, I did a little Irish jig in the middle of the room in celebration. There were hugs and smiles and everybody felt they were going to change the world."

"This particular device has various interested parties involved, and work is still ongoing to establish a route to market."

More successes followed.

"The floodgates were opening; it seemed something had shifted in the ether. I was next invited to see a device in the South of the United States. An elderly inventor, who had developed his over-unity motor concept in his basement over a decade with virtually zero funding, revealed to me another working device with a CoP of 3, meaning three times more power out than power in."

ALMA MATER MOCKS INVENTOR

The inventor had spent much of his life working at a prestigious American university. After retirement he went back to the university to reveal his astonishing breakthrough in motor technology that could allow electric cars and airplanes to run without fuel.

At the university, the inventor was mocked and not listened to. He told John that the academic specialists would not even use their oscilloscope and measure the power of the device. "This is a typical response from mainstream academia to these concepts," John said, "and the inventor is still working in his basement."

SELF-LOOPED DEVICE

Another device in the United States soon appeared on his schedule for testing. It seemed to be the holy grail of machines, called a self-looped device because of ability to sustain itself. Part of the power coming out from the machine is looped back to the machine's input so that the system can operate without external input power.

Except for maintenance due to wear and tear such as worn bearings, that theoretically means the device should run without stopping—the long-derided "perpetual motion machine." Despite the supposed impossibility, in the demonstration lab of the small company John witnessed a self-looped machine, without any external wires or cables, powering many kilowatts of load.

'WATER IN, ELECTRICITY OUT.'

Within two weeks of that self-looped demonstration, Susan and John were in the laboratory of another small American company. They witnessed what he describes as convincing proof of a machine that creates clean and abundant amounts of electricity based only on water as an input.

According to the inventor, scientists misunderstand the fundamental nature of hydrogen. He said that with knowledge about how to treat water with certain frequencies, hydrogen can be chemically unbonded from the oxygen in the water by using no electrical power apart from a "tickle" of a certain frequency. John had heard this from other inventors as well. The ability to create significant amounts of hydrogen nearly free of cost would alone solve the world's energy problems.

Since John Cliss is not a chemist, he cannot comment on the possibility that the American inventor's method was valid. "But when I read third-party independent reports from professional engineering companies concluding 'water in, electricity out,' it is difficult to deny that it works."

Excited about what might be achieved with the technologies he had seen, John returned to Europe in May of 2018.

> **It was yet another concept resulting in seemingly endless supplies of clean and free electricity.**

Through coincidences, he ended up in a hotel room in Southern Europe and witnessed yet another self-looped machine. This one was about the size of a washing machine and powering several bedside lamps. A brief electrical jolt from a battery started the machine and allowed it to increase speed up to a certain number of revolutions per minute. Then the operator hit a switch, removed the battery, and the device powered itself. It kept the lamps fully lit for the full twenty minutes of the demonstration.

John was allowed to inspect inside the device and saw no space that could hold hidden batteries. He was satisfied that the machine was doing what the inventor claimed. In this case, the machine used the power of centrifugal force. It was yet another concept that could result in nearly endless supplies of clean and free electricity.

He recommends that any potential investor in these technologies arrange for stringent testing[39] by a series of well-controlled experiments. A third-party engineering company with no financial interest in it would assess the device, as has been done in several of the examples he describes.

Susan Manewich, Energy Science & Technology Conference

After John had time to digest the glut of new technology that had appeared in his life, he and Susan tried to get funding to the inventors. However, they concluded that solutions to humankind's energy problems were not found in merely chasing technologies, but in creating an organization, with ethics and integrity, that would be a safe container for a new paradigm to be birthed into.

---

39   The device should be taken to a neutral setting with a controlled environment to remove possibilities for hidden batteries or cables, or induction coils hidden in the ground etc. John Cliss says the machine should be self-looped and run for a significant length of time. Cliss has seen third-party tests ranging from five days to six months. After testing, the third-party engineers should completely deconstruct the device, then reassemble it and start again. If the device works, then this would be adequate proof that such devices work. Cliss personally had not yet had the opportunity to conduct such extensive tests.

CHAPTER 6
THEY SEE A BETTER FUTURE

Building factories and zero-point energy machines would provide more than enough jobs to employ everyone displaced by the abandoning of oil.

—Hon. Paul Hellyer, author, statesman.[40]

THE IMAGINATIONS OF DESIGNERS AND BUILDERS ARE limited by what they are taught about how things work. Engineers won't invent a car or power plant that uses a truly new source of energy if they never envision such a source.

Decision-makers who plan cities and towns, job creation projects, health-care, agriculture and many sectors may be unaware of possibilities for a global Renaissance in which technology works in harmony with nature. However, John Cliss and others in this book have given much thought to how energy breakthroughs may change our world. "Let your imagination be the limit," he says, and predicts some aspects:

In the first stage of change, cars, trucks, boats and planes will be made electric by using new energy devices. That removes pollution from travel. People will be able to work more remotely if they wish to, away from crammed cities. The new energy devices will power, heat, and cool buildings—again minimizing pollution. There will be no need for vast electric power grids.

---

40   Hellyer, Paul, *Hope Restored,* TrineDay, Oregon, 2018.

END GLOBAL AGGRESSION, FOCUS ON WATER AND FOOD

Wars of aggression seeking control over oil producing countries will end, he predicts. This will radically change the political landscape.

Fresh water will be seen for what it is, vastly more valuable than oil. New energy technologies will not only put an end to fuel spills, they can provide power for cleaning polluted waterways. Widespread recognition of the priceless nature of water is crucial to humankind's survival.

Although there may still be tension over access to certain raw materials, the theories behind "free energy" devices lead to speculation that replicator technologies will be able to create any type of substance including metals, Cliss said.

Materials might not even need to be mined. Dr. Randell Mills (Chapter 20) saw the byproduct of his invention combine with other elements and create new compounds. There could be thousands of novel compounds with different properties, so why not new building materials?

People could feed themselves easier if they had extremely low-cost electrical greenhouse heating, or mechanical energy for farm equipment. While the need to shift from polluting to breakthrough energy technology is urgent, so is a shift to regenerative agriculture. Those methods grow food without needing the petrochemical-based fertilizers that kill soil organisms and ecosystems.

BENEFICIAL AND DANGEROUS POSSIBILITIES

What about space travel?

"After electrification of our vehicles, we will see new designs of air vehicles that will get us into space to explore the solar system and beyond. This has enormous ramifications—practically, spiritually, morally, philosophically."

In contrast to his enthusiasm for new energy possibilities that could expand humankind's horizons, Cliss warns about a different aspect—weapons that could be produced from the technologies. He added that governments or groups controlling governments have had access to exotic technologies for decades.

"We now need to bring an understanding of these technologies to the people of the world so that we can identify their misuse and make laws to ban their operation for the uses of war and aggression."

THE WORLD ECONOMY

Release of new technologies may cause economic problems if they replace fuels too quickly. It is important to lessen that impact. John Cliss compares the process to clearing toxins out of the body; people who do an internal cleanse for health may feel worse before they feel better. The endless-war economy has been a source of toxicity for society.

Those who benefit from that economic system will at first try to control the new inventions, Cliss said. He believes they won't be able to stop the shift to decentralized power. His hope is that any economic upheaval will relax into a future founded on a fair economic system.

HEALTH

People may worry that unlimited electrical energy will increase the ill health from electro-smog, Cliss expects. He sees these concerns as real. However, planners can make wiser choices than are made today.

Currently we are in an electro-smog environment without understanding what causes the harmful impacts on health, he says. "Why do mobile phones cause cancer? I have studied much in the free energy theory that relates directly to why these negative health effects occur and how it can be possible to develop electrical technologies that are healthful... So this (beneficial knowledge) will bring huge medical breakthroughs as well."

"I think the breakthrough in our theoretical understanding of how free-energy systems work will also lead to related breakthroughs, for example, in understanding how some electromagnetic frequencies we are exposed to from electrical devices can make us sick whilst some can make us healthy."

TOWARD PEACE

He expects the release of free energy devices to be a crucial step toward world peace.

"...Pollution-free technology to power our buildings and power transport, combined with harmonious agriculture, adventure into space, reduced wars, cleaner politics, a fair economic society and good health will significantly reduce the stressed state that we all experience in everyday life in the 'rat race.' I believe this will lead to a more harmonious society that will begin to really tap into the true creative power of the human being."

COMPARING PARADIGMS

To view possible outcomes of a societal shift from energy-scarcity fear to abundance-thinking, we made a simple chart. It compares the materialist worldview that dominates today, versus values that may inspire tomorrow's worldview:

Extremes of old paradigm	Ideals of new paradigm
Materialism: 'Only matter matters.'	Sacred science: Nature's beautiful creativity.
'Net worth' of person stated in dollars.	Integrity, ethics, altruism, love are valued.
Believing life is purposeless, random.	Life has purpose, order *and* disorder.
Old Boys' Club attitudes prevail.	Balance: feminine and masculine leadership.

Extremes of old paradigm

Ideals of new paradigm

Extremes of old paradigm	Ideals of new paradigm
Engineered man is a machine with AI.	Human spirit connected with Divine.
Centralized, vast electricity grid.	Small local or regional clean-power systems.
Polluting energy technologies.	Fuel-less energy systems, no negative impact.
Scarcity of energy resources.	Abundance of clean energy responsibly used.
Huge military and surveillance budgets.	Life-enhancing job creation.
Corporate control of resources.	Bioregional and local stewardship.
Environment trashed.	Regeneration of ecosystems.
Throwaway culture.	Cradle-to-grave sustainability.
Monoculture mentality.	Diversity encouraged.
Cutthroat competition.	Cooperation, collaboration.
Consumer society.	Relationships, arts, beauty, nature valued.
Alienation.	Awareness of connection to all life forms.
Centralized governance.	Distributed authority.
Injustices are ignored.	Justice and opportunities for everyone.

You may have a different set of old- vs. new-paradigm comparisons.

We present these extremes as a discussion-starter. You and your circle can make your own list in the spirit of "what we envision we co-create."

CHAPTER 7
WHY NOT GIVE IT AWAY?

The problem we have here is consciousness, red tape, and the power our current energy industry has over the global economy.

—Arjun Walla, Collective Evolution[41]

INVENTORS WORKING ON REVOLUTIONARY ENERGY- converting technologies face extra challenges.

Often an inventor spends his or her own money on materials, electronics and machining for the early-stage experiments, then borrows from family and credit cards to fund the next stage. Each working prototype needs to be engineered into a product certified as safe for you to buy. That's the most expensive stage, often requiring a team that includes expertise the inventor lacks.

Inventors are also faced with the daunting expense of hiring a patent lawyer. If they don't protect their inventions with patents, they don't attract the investors with "deep pockets"—access to enough money to propel an invention into the marketplace.

---

41   Walla, Arjun, https://www.collective-evolution.com, Nov. 13, 2018.

'WHY DON'T THEY OPEN SOURCE?'

The praise-worthy spirit of open sourcing is strong in today's younger genera-tions, due to the success of developing software that way. Open sourcing means sharing the technical details and allowing anyone to build the thing for themselves as famous entrepreneur Elon Musk recently announced he would do. He is giving away information from his company's electric vehicle patenting. In the software field, the original team or inventor open sourced the software then received feedback on how to improve it.

Open sourcing can tap into the brain power of thousands, instead of keeping details secret until a product is fully developed. One example in the new energy field is The Martin Fleischmann Memorial Project, a group who experiment in the field formerly called cold fusion and now referred to as LENR (see Chapter 20).[42] The project publicly shares its procedures, data, and results online.

Inventions have disappeared when the originator died, so why don't all inventors of energy devices just give their intellectual property to the world?

Jason Verbelli is involved in breakthrough energy work and is of the genera-tion most enthusiastic about open-sourcing. He explains that thieves and unscrupulous people are why open-sourcing is not always the best route for getting energy inventions to the people. "Let's say you spend one million dollars out of your own pocket to experiment for a few years, purchase materials and equipment, pay for utilities, etc. Someone will take that free information you paid for. They will develop their own version, patent it, form a company and then sell production units."

"And if others patent the information before you, they can sue you and prevent you from continuing your own project. Then you have no way to recoup the money that you paid out of your own pocket to develop everything. They could sell the product to another company which might just put it on the shelf and it would never see the light of day."

Verbelli is on a research team supportive of the elderly inventor John Searl, who has for decades open-sourced information about his innovation. However, Searl keeps a vital part of his Searl Effect Generator (SEG), the

---

42   Low Energy Nuclear Reactions.

magnetization process, a trade-secret. Searl likens it to Coca Cola's recipe, which is neither patented nor public knowledge yet the company functions internationally. Verbelli explains that having a secret recipe makes the company viable.

Even with thousands of pages about Searl's generator freely available for decades, few experimenters ever tried to build their own test platform or SEG. Verbelli asks rhetorically, "Why provide proprietary information to people who don't display genuine interest with everything that is already available?"

Some projects should be open sourced, Verbelli advises inventors, but "don't sabotage yourself and everyone else by releasing information too early because of noble intentions and a big heart. Smart planning is just as important. Otherwise, the seemingly noble actions you do to 'save' things might actually cause their downfall. Healthy balance is the key."

That prevents someone from patenting the invention and then shelving it.

BLIND CONFUSION

One of the problems for new energy science is the lack of agreement on terminology. It reminds us of the fable in which a dozen or so sightless people spread out around, under and on an elephant. Each described it by touching the part near them. The one who grabbed the tail said the elephant is a rope. The man with his arms around a leg said the elephant is a type of tree trunk. The person standing near the elephant's belly insisted the object was neither a rope nor a trunk.

People don't like ambiguity, but until the emerging science about universal energy matures, we need patience with the lack of agreement on what it is.

A concept does need a name everyone can agree on. If mainstream thought leaders, mass media, and those who write textbooks ignore the concept, then anyone who does notice its existence and wants to talk about it can create a name. That's a recipe for chaos, and chaos is what often happens. A humorous incident illustrates that.

CONFUSED ENTREPRENEUR

A man who looked to be in his twenties—we'll call him Alvin to protect his identity—showed up at an International Tesla Society conference. He advised people that he was a non-technical guy but knew "people with big money." He wanted to meet inventors of market-ready breakthroughs and connect them with funding. As the go-between, he would collect a finder's fee. After listening to two days of technical talk, however, two facts struck him:

If anyone had developed a revolutionary energy generator to the commercial level, ready to take to the marketplace within a year, that inventor had no need to attend conferences and had stayed home. The other hard fact was that if Alvin were to be a go-between, he would have to speak technical language or at least be able to understand conversations. Maybe he would even learn what "quantum fluctuations of the fabric of space" means.

At the social hour on the last night of the conference, he admitted that his brain was overloaded with unfamiliar physics terms. "Zero-point quantum field. The fabric of space. Spacetime. What the f--- *is* spacetime, anyway?"

Since his head already felt like it was spinning, he decided to toss a few alcoholic drinks down his throat. The social hour stretched into hours and his amusement progressed from chuckles to knee-slapping laughter. "Energy from the vacuum... I guess that doesn't mean a Hoover..."

Alvin was puzzling over the scientists' references to "the fabric of spacetime" when his words became slurred. His final contribution to the evening was delivered loudly as he waved an empty wine glass.

"The fabric of space is...polyester!"

TOO MANY NAMES

The words on our graphic in the Introduction are less imaginative but more accurate, ranging from torsion fields (Russian scientists' term, also called spin fields, referring to the spin of particles) to life force[43] or zero-point energy

---

43   *Life Force, the scientific basis,* by Claude Swanson, PhD, contains ample experimental evidence of such a force.

(from quantum mechanics) and Dirac Sea (named for a Nobel laureate) to orgone (Dr. Wilhelm Reich) or Radiant Energy (Nikola Tesla's term) and The Field.[44] The field of breakthrough energy inventions needs an official professional association of its own, to settle its terminology differences.

---

44   McTaggart, Lynne, *The Field*, HarperCollins UK, 2001.

PART II
HIDDEN ENERGY

CHAPTER 8
TESLA, THE MAN
AND THE TOWER

"Ere many generations pass, our machinery will be driven
by a power obtainable at any point in the universe...
Throughout space there is energy."

—Nikola Tesla

THE PERSON ON THE STREET LIGHTS UP WITH RECOGNI-
tion at the name "Tesla." However, most people don't associate the word with the man—inventor Nikola Tesla. They usually learn in school that Thomas Alva Edison invented the light bulb but are not taught that Tesla lit it up. Nikola Tesla was a world celebrity in the Edison era.

More than a century after Nikola Tesla's height of fame, American entrepreneur Elon Musk made "Tesla" a household word again. Musk built and marketed an advanced electric car after he bought a company that had chosen the name Tesla Motors.

The car was a hit. Musk in 2014 broke ground in Nevada for a "Gigafactory" to produce Tesla electric motors and battery packs. Another venture, Hyperloop, was intended to move people through tubes at airline speeds for the price of a bus ticket. And he proposes to solve traffic jams by drilling layers of transportation tunnels under major cities. Musk founded yet another company, SpaceX, with plans for taking people to Mars. One of his rocket launches hurled a red Tesla Roadster into orbit around the sun at unbelievable miles per hour.

Which makes us wonder what frontiers an Elon Musk will open on Earth ten years from now. We can only speculate about how far Musk might advance beyond the conventional technology he's innovatively using. In the Tesla car, Musk uses Nikola Tesla's 19th century invention of an alternating current induction motor. Will Elon Musk eventually incorporate revolutionary energy secrets that Tesla, the man, discovered later in life but are mostly unknown today?

Musk has said he is philosophically closer to Nikola Tesla's old rival, Thomas Edison.[45] Musk and Edison each focused on building businesses. Nikola Tesla didn't.

Edison commercialized his own and others' inventions. In his Menlo Park, New Jersey, industrial laboratory, his company improved on others' patents and made money on applications. When he died, his fortune in 1931 dollars was equivalent to billions of today's dollars. Edison Electric Light Company became General Electric Corporation, which, we are told, quietly uses some of Nikola Tesla's inventions.

NIKOLA, ON THE OTHER HAND...

Nikola Tesla wasn't a marketing/financing/billionaire type, but he did think big. He set lofty goals for the good of humankind, as Elon Musk reportedly does.

Nikola Tesla invented basics of our everyday use of electricity, plus radio and many other helpful devices we use. Strangely, Edison often gets more credit for the electrical grid than the inventor who had a larger role in developing alternating-current (AC) systems, Tesla. What Edison built in 1882 was limited—a generating station that distributed direct current (DC) to less than 100 customers in lower Manhattan. The station couldn't serve customers farther away. If Edison had been using AC, he could have had

---

45   Musk, Elon, interviewed for The Henry Ford's Visionaries on Innovation series, 2008.

electrical transformers along the way, restoring the voltage levels. That doesn't work with DC.

Nikola Tesla had decided to solve those limitations, even before he emigrated to America. When he told his European professor about his plan to invent a new form of electric motor that uses alternating current,[46] the professor dismissed it as impossible. Later, in Budapest, Hungary, Tesla experienced a vision that revealed how to do it. After immigrating, he created and patented the many parts of the AC system.

BACKSTORY TO INCIDENT IN BUDAPEST

Nikola Tesla's extraordinary abilities may explain how he could create something that an expert such as his professor insisted is impossible.

He was born in 1856 in an eastern European region then called Croatia. His father was a Serbian minister and his mother an inventor. Tesla's sensitivity and illness as a youth forced him to look inward and control his thinking. Later, he expressed gratitude that those difficulties taught him how to focus. He also had what are known today as out of body experiences—journeying to new countries and making friends there, but not in his physical body.[47]

While at engineering school, Tesla concentrated so intensely that he was hit with what his doctors called a nervous breakdown. His vision and hearing became more painfully sharp than ever. He said he could see any invention he was thinking about, as clearly as if the device were solidly in his hands. Or he perceived it like a detailed hologram that he could walk around.

To restore his health, he often hiked with a friend. One afternoon they strolled through the Budapest city park during a luminous sunset. Tesla spontaneously recited the lines of a poem by Johann Wolfgang von Goethe:

The glow retreats, done is the day of toil;

It yonder hastes, new fields of life exploring;

---

46   Nikola Tesla's AC *induction motor* means that its stator windings induce a current flow in its rotor conductors. It is like a transformer but unlike the DC commutator motor which relies on brushes to make electrical contact.

47   Seifer, Marc J., *The Life and Times of Nikola Tesla: Biography of a Genius*, Citadel Press, 1998, p. 11.

Ah, that no wing can lift me from the soil,

Upon its track to follow, follow soaring…

Suddenly he was spellbound by the setting sun and a vision of a vortex whirling in space. The answer to the technical challenge that his professor had said was impossible came like a lightning flash. Mystics describe such an experience as an opening of inner vision.[48]

That perception, beyond ordinary sight, revealed to Tesla an elegantly simple motor operating in front of him. It was a completely new approach, a rotating magnetic field producing alternating electrical currents that are out of phase with each other. He saw a workable system that would revolutionize electrical technology.

He tried to convey his excitement, "See my motor… Watch me reverse it!"[49] His friend couldn't see the vision and must have thought Tesla had lost his mind.

Tesla grabbed a stick and drew a diagram in the dust. Six years later, he would show a similar diagram to the American Institute of Electrical Engineers and tell the world about his new scientific principle involving a magnetic whirlwind created by out-of-step electrical currents.

OFF TO AMERICA

In Paris, Tesla got a job as an engineer at Continental Edison. Thomas Edison was already famous. When Tesla decided to immigrate, his boss wrote a note to introduce him to Edison, "I know two great men and you are one of them; the other is this young man."

Tesla left by ship for America in 1884. Soon after Edison hired him as an assistant, Tesla won Edison's grudging respect for working eighteen-hour days, seven days a week and conquering technical problems. However, Edison eventually lost his industrious new assistant.

Tesla had described how he could improve the efficiency of Edison's dynamos. According to Tesla, Edison had replied, "There's 50 thousand

---

48   The mystical or esoteric term for that ability is 'opening of the Third Eye.'

49   Cheney, Margaret, *Tesla: Man Out of Time,* Bantam Doubleday Dell NY p. 23.

dollars in it for you if you can do it." Tesla did successfully redesign Edison's machines after months of intense work.

The immigrant was shocked when he asked for his promised bonus and Edison replied, "Tesla, you don't understand our American humor."[50]

Edison wouldn't pay. Tesla walked out. He made a living digging ditches on a New York street crew. Three years later, his luck changed; financial supporters gave him the chance to have a laboratory and develop his system of alternating current.

WESTINGHOUSE PAYS; EDISON FIGHTS

Tesla could have been very wealthy. The industrialist George Westinghouse bought Tesla's forty AC patents and signed a contract to pay Tesla royalties based on horsepower of his AC products sold, expressed in watts.

Edison fought the introduction of alternating current electricity because his lamps ran on direct current (DC). During what became called the War of the Currents, Edison electrocuted dogs and published scare pamphlets, all in a public relations attempt to link alternating currents with death. Tesla and Westinghouse won anyway.

Tesla's system lit the 1893 World's Fair in Chicago, and the handsome Serb was a star at the exhibition. In white tie and formal jacket with coat tails and wearing cork bottomed shoes for protection from electrical currents, he shared a stage with one of his Tesla coils, a device he invented that generated high power currents. Its bolts of electricity crackled and snapped and lit lightbulbs in his hands. Crowds loved the drama. The exhibition's success led to development of a hydroelectric project at Niagara Falls.

---

50   Ibid. p. 33.

TESLA RIPS UP CONTRACT

Tesla should have been financially wealthy for life, since the contract with Westinghouse gave Tesla $2.50-per-watt royalties. It didn't happen that way. Instead, George Westinghouse fell into financial troubles. Investment bankers told him to get rid of his generous contract with Tesla. When Westinghouse appealed to Tesla, the inventor remembered that the industrialist had believed in him when no one else had.

Tesla tore up the contract, took a cash settlement and walked away from the fortune assigned. Although he enjoyed money when he had it, he wanted to see his friend's company survive and ensure that the AC system would be available to the world.

CELEBRITY

Tesla fascinated New York society. Always impeccably dressed, slim and tall, he commanded attention with aristocratic bearing. Intense blue eyes and his aura of mystery attracted women of wealth and culture who developed romantic crushes on the celebrated inventor. He enjoyed their dinner parties, yet his mission to help the world was his priority.

He mesmerized guests at his laboratory in New York City by demonstrating electrical effects. His intellect attracted cultural stars including author Rudyard Kipling, architect Stanford White, pianist Ignace Paderewski and naturalist John Muir. Samuel Clemens, well known as the humorist/author Mark Twain, was his close friend.

A French stage actress figures in an aspect of the Tesla story told in the next chapter. On one of her American tours, Sarah Bernhardt acted in a play about the life of Buddha. When she saw Hindu religious teacher Swami Vivekananda in the audience, she arranged a meeting which included her friend Nikola Tesla. After that the inventor showed up on other occasions to learn from Vivekananda about Eastern science.[51]

ONWARD TO WORLDWIDE BROADCASTING—OR NOT

After the 1893 World's Fair, Tesla gave a speech at a scientific institute in Philadelphia and talked about his vision for his next big invention which he called a magnifying transformer. In the next few years Tesla patented processes for its wireless transmission of power and messages.

It was a futuristic plan. Power would reach a receiving station halfway around the earth without any connecting wire. Unlike today's wireless technologies that transmit through the air, his wireless involved sending out electrical pulses of a specific frequency in one direction through the earth.

The earth resonance part of his wireless plan is easy to understand if you're a musician. Just as a piano string will vibrate when another instrument at a distance hits the same note as the pitch to which the string is tuned, a distant Tesla receiver would be tuned to resonate with whatever the Tesla transmitter broadcasts.

He had a laboratory built in the mountains of Colorado to test his ideas about wireless, and afterward told the Los Angeles Times, "With my transmitter I actually sent electrical vibrations around the world and received them again and I then went on to develop my machinery."

His announcements suggested that his Magnifying Transmitter would transform both power generation technology and the way that electric power is sent. He predicted that such a system someday would also speed communications. Imagine a satellite system without satellites. It could also have been an alternative to the inefficient and vulnerable old web of power lines blighting our landscapes.

---

51   "The Influence of Vedic Philosophy on Nikola Tesla's Understanding of Free Energy" by Toby Grotz. Web Publication by Mountain Man Graphics, Australia - Southern Autumn of 1997.

After his experiments at Colorado Springs, Tesla turned toward Long Island, New York, where he began building a large facility called Wardenclyffe. A sphere of more than 60 feet in diameter topped his structure. Tesla claimed to be able to charge it to 30 million volts by a simple device that provided static electricity and power. His system was not a generator of electrical power but rather it was a receiver intended to amplify the power collected.

The banking mogul J. Pierpont Morgan funded the Wardenclyffe project for a while, believing that Tesla only intended it for communications across the world. Tesla eventually revealed to Morgan his power for the people plan—a system that can't be metered and billed.

That did not go over well with industrialists who had already bought up copper mines after predicting that utility grids would cover much of the world with nets of copper strands. Tesla's funding dried up, and construction of the Wardenclyffe transmission tower stopped. Further ruining his chances to complete that tower, World War I created fears about communicating with an enemy. The tower was demolished for scrap in 1917 to pay Tesla's debts.

Tesla soon fell from fame. Not only was he blackballed by the banking/railroad mogul, he was an outsider to other sectors. Fueling criticism, he used newspapers to propose scientific developments which were beyond his ability to prove since he no longer had funding. His unsupported announcements at annual press conferences fed the part of the scientific community which wanted to dismiss him as a 'crank' and belittle his earlier accomplishments.[52]

Would-be debunkers ignore the work that Tesla continued for decades early in the 20th century. He still found ways to experiment and discover the strange cold type of electricity that he called Radiant Energy.

MAGNIFYING POWER FOR ELECTRIC CAR?

Energy historian Dr. Peter Lindemann commented on mysterious later years of Tesla's life. "After Tesla was prevented from bringing his World Broadcast System into full manifestation, he worked for years to develop a smaller

---

52   Terbo, Wm. H., (Nikola Tesla's great-nephew), "Opening remarks" in S.Elswick, ed., *Proceedings of the 1988 Tesla Symposium*, Colorado Springs, C , 1988, p. 10.

version of the device that harnessed the same principles. By the 1920's he had succeeded. This specialized electronic circuit is what powered his infamous Pierce-Arrow automobile."[53]

The electric Pierce-Arrow legend is controversial. Researchers are unable to locate physical evidence of the car. Also, previously the story had been told by only one man who claimed to have been a witness. That man referred to Nikola as "uncle" but could not have been a blood relative.

New evidence regarding the electric automobile story surfaced recently, although it only verified that a Heinrich Jebens, who headed a German institution for inventions, had visited Edison in the time period Jebens claimed in a confidential memo.

When Heinrich Jebens died, his son Klaus found the secret memo Heinrich had written to himself about his visit to America to give Edison an award. The German official wrote about a shipboard meeting with someone who, after they arrived in the USA, connected him with Tesla. Jebens' memo said the famed inventor had given him a ride, under strict secrecy, in a car powered by a device resulting from one of Tesla's "earlier patents with ether energy." The Pierce-Arrow car's gasoline engine and fuel tank had been removed.[54]

Thomas Valone, PhD, author of technical books and anthology on Tesla's wireless.
(Jeane Manning photo)

---

53   free-energy.ws.

54   A History Channel researcher dug into Thomas Edison's archives for a television series about Tesla, and confirmed that Heinrich Jebens had visited Tesla in the time period claimed in a book by his son Klaus Jebens. Our article about Heinrich's report on Tesla's Pierce-Arrow sedan is on the HiddenEnergy.org website. Klaus Jebens' book is *Die Urkraft aus dem Universum* Jupiter-Verlag, Zurich 2006, ISBN 3-906571-23-8.

Tesla fans today are excited about a "Tesla tower" recently built in Texas to send power along the ground. If fans want technical history of it, the anthology *Harnessing the Wheelwork of Nature*, edited by Dr. Thomas Valone, contains papers about Tesla's energy science. Electrical engineer James F. Corum, PhD, and his brother, experimental physicist Kenneth L. Corum, are included; they patented aspects of what is said to be Tesla's wireless power. Dr. Corum is now chief scientist for Texzon Technologies, which built the tower near Milford, Texas. The company describes itself as pioneering electromagnetic wave propagation, power storage, and electricity distribution.

Dr. Valone's anthology includes a paper, "Harnessing the Earth-Ionosphere Cavity for Wireless Transmission," by well-known physicist Elizabeth Rauscher, PhD, and the late William Van Bise. Dr. Rauscher spoke at the Integrity Institute's Wardenclyffe Tower Centennial in 2003. Dr. Valone wrote about her contribution, "Unknown to most electrical engineers, Tesla's dream of wireless energy transmission provides a real alternative to transmission lines..."

"Dr. Rauscher indicates that the earth's ionosphere and magnetosphere are a source of electrical energy, as Tesla emphasized, triggered by the relatively small longitudinal wave impulses that the resonant Tesla Wardenclyffe Tower supplied. Dr. Rauscher shows that the available power of the earth-ionosphere cavity is close to three terawatts (three billion kilowatts), while the U.S. only consumes about 425 million kilowatts today for electrical needs (at 26 per cent of the world usage at the time)."[55]

TESLA TOWER MAKING WAVES

Recent talk about that "Tesla tower" centers on waves, disturbances that move through space and time. A *longitudinal* wave is made up of compressions and rarefactions (areas of reduced density) while a *transverse* wave is made up of crests and troughs.

---

55   Valone, Thomas, editor, *Harnessing the Wheelwork of Nature*, Adventures Unlimited Press, 2002, p. 238-9.

To picture transverse waves, hold one end of a rope that's tied to something else. Move the rope up and down. Imagine the waves traveling perpendicular to that movement. To picture longitudinal, think of sound waves coming from a loudspeaker. They oscillate in the direction the sound is headed. The sound alternates between compressing (bunching up) and rarifying (stretching forward).

Tesla had insisted that his system for sending electricity was the opposite of a Hertzian (radio wave) system which radiates out all the time regardless of whether the energy is received or not. One theorist put it simply, "Disturbance in the ether is longitudinal; Tesla was right."[56]

In Tesla's system, when a tuned, resonant receiver was turned on, it drew power. However, when there was no receiver on there was no energy consumed or dissipated.

The Corum brothers, however, deal with waves that go along the surface of the Earth, called Zenneck waves. Long-time researchers say that Tesla sent oscillations into the earth, not just along the surface.

OTHER RESEARCHERS WEIGH IN

Many inventors study Tesla's Colorado Springs Notes and the magnifying transmitter. Some also discover the "cold electricity" that Tesla called Radiant Energy and that seems to be a safer form of power. It illuminates light bulbs but must be converted to a standard form of current before running regular appliances. Charging batteries with it is one way of converting cold electricity to useful DC.

Electrical engineer Eric Dollard is an expert on lost or forgotten electrical science and is said to have successfully replicated Nikola Tesla's magnifying transformer, decades ago.[57] Dollard seeks funding to resurrect ambitious projects. The Zenneck wave system has nothing to do with Tesla, Dollard

---

56   Thomas Bearden
57   Borderland Sciences Research Foundation took videos of Dollard's earlier experiments and published his articles. His recent books and lectures are available through eMediaPress.com.

insists. "The Zenneck wave is an electromagnetic form of wave. Telsa said, again and again and again, that his wave was not electromagnetic."[58]

Dollard's view is independently voiced by others who study the ether (see Chapter 10) aspect of Tesla's work. Norman Wootan, for instance, has pored over Tesla's Colorado Springs notes. Before retirement, Wooten had a career working with electrical and electronics equipment for his country's military and industry.

Wootan's friends know that his travels take him near Milford, Texas, and past the strange tower with the bulbous top glimpsed from the highway. "Why don't you stop and investigate that Tesla tower?" they ask. "Why should I?" he replies. The Milford project is all about proprietary technology; he wouldn't want to trespass. "And what they're doing isn't what Tesla was into."

Wootan explains that Tesla's Wardenclyffe Tower and his experiments at Colorado Springs were to verify Tesla's different vision and methods for earth-transmitted energy.

WHAT NON-TECHIES WANT TO KNOW

The Texan tower or further structures from the same companies won't fulfill Tesla's dream of free power to the people. Instead, the companies have encryption and other technologies to ensure that no one can hack into their signals. Their plan is to transmit waves that only their substations in Bahrain or elsewhere in the world can convert into electricity. The substations will then dole out the electricity through a standard power grid. You still must pay a utility bill.

However, if it works when scaled up, a new way of delivering electrical energy wirelessly over long distances could send power to remote or disadvantaged areas and eliminate long-distance wires and cables.

A mindful approach to a new technology is to ask questions that its promoters might not bring up. Could use of Tesla wireless be beneficial or harmful to living beings depending on what frequencies are transmitted? If the scientists

---

58   Dollard, Eric, on public calls hosted by Aaron Murakami, February 2019 and March 17, 2019.

involved are ensuring that their technology is at least benign, they will be investigating its magnetic or electric field effects on life forms.

Dr. Rauscher and Van Bise were ahead of their time with their concerns in 1997 when their wireless transmission paper cited was first published.[59] They concluded the paper with, "The earth and the life forms upon its surface vibrate and resonate in harmony in such a manner that radiant energy from the sun and materials and vibrations of the earth support this life....Also impressed upon the environment are many man-made sources disturbing both the atmosphere and the earth.... We must examine what we are doing as people, as societies and as nations! If we do not develop a new consciousness and awareness, destruction of life will inevitably result."

---

59   Rauscher, Elizabeth, and Van Bise, William paper "Harnessing the Earth-Ionosphere Cavity for Wireless Transmission," reprinted from 1997 *Tesla: A Journal of Modern Science*.

CHAPTER 9
THE VEDIC CONNECTION

Tewari's discovery requires that the laws of physics, as now taught, must be modified to recognize that space is not empty, and the substance of space is the origin of matter—as known by the ancient seers of India for more than 10,000 years.

—Toby Grotz, electrical engineer[60]

DR. ELIZABETH RAUSCHER AND ELECTRICAL ENGINEER Toby Grotz sent out a call for scientific papers about Nikola Tesla's work, and she convened a Tesla Centennial symposium in Colorado Springs in 1984. Several hundred engineers and others then celebrated the centennial of Tesla's arrival in America.

The International Tesla Society was founded at that symposium.[61] As its first president, Grotz chaired the next symposium, where Jeane met him. We chronicle his search for knowledge because he encountered fundamentals of universal energy. And he met an extraordinary engineer in India, Paramahamsa Tewari.

Toby Grotz' search began as a young man, after a herniated disc resulted in a spinal fusion. He had been injured during work on a seismic exploration

---

60   The *Upanishads* are the final part of the *Vedas*—the ancient and sacred scripture of India. The Indian sages wrote the *Upanishads* between the eighth and fourth centuries BC.
61   New York Times, Science Times section, August 28, 1984. Syndicated copies of the report about the symposium appeared in newspapers around the world, including the International Herald Tribune.

vessel. That was in the 1960s when young people dove into esoteric teachings, ancient cultures, and meditation. To care for his back, he began a lifelong practice of hatha yoga. The book *Autobiography of a Yogi*[62] further sparked his interest in Hindu philosophy. He also began the study and practice of Eckankar which at that time was subtitled the Ancient Science of Soul Travel.[63]

Grotz took classes from a scholar[64] of Sanskrit, the language of the Vedas. Vedic writings date back at least 5,000 years and range from atomic structure to ancient advanced civilizations. His friend, Vedic scholar Robert E. Cox, later wrote *The Pillar of Celestial Fire*. Vedic science recognizes a field of primordial energy called Prakriti which Cox said is "unmanifest."[65]

According to the Vedas, a 13,000-year halftime on the "cosmic clock" of the zodiac is ending and our chaotic times can be the entry to a golden age of harmony. Cox predicted miraculous new technologies will develop from this era of great change.

When Grotz read the Tesla biography *Man Out of Time,* by Margaret Cheney, he noticed Tesla used Sanskrit words such as *akasha* and *prana*. Prana means energy, usually translated as life force, and akasha is the ether. Grotz wondered if those terms are a key to Nikola Tesla's view of electromagnetism and the universe. In a book by Tesla's friend John J. O'Neal,[66] Grotz found an illuminating excerpt from Tesla's 1907 article "Man's Greatest Achievement."[67]

Tesla wrote, "Long ago (human beings) recognized that all perceptible matter comes from a primary substance, or tenuity beyond conception, filling all space—the Akasha or luminiferous ether which is acted upon by the life-giving Prana or creative force, calling into existence in never ending cycles all things and phenomena. The primary substance, thrown into infinitesimal

---

62   Yogananda, Parmahansa, *Autobiography of a Yogi.*

63   Eckankar, The Path of Spiritual Freedom. www.eckankar.org.

64   Neal Delmonico taught Sanskrit at Denver Free University. He had been secretary to A.C. Bhaktivedanta Swami Prabhupada, founder of the International Society of Krishna Consciousness.

65   Cox, Robert E., *The Pillar of Celestial Fire and the Lost Science of the Ancient Seers*, Sunstar Publications, Iowa, 1997, preprint version.

66   John J. O'Neal Prodigal Genius: The Life of Nikola Tesla Cosimo Classics NY, 2006.

67   Tesla, Nikola, "Man's Greatest Achievement," written in 1907 and published in New York American - July 6, 1930.

whirls of prodigious velocity, becomes gross matter; the force subsiding, the motion ceases and matter disappears, reverting to the primary substance."

Tenuity means low density or thinness; "beyond conception" could mean nearly total lack of matter. Tesla not only investigated the Eastern view of what drives the material world, he wrote to mathematical physicist Lord Kelvin about it.[68]

Tesla's insights may have come from his own experiences. Grotz speculated that, in the Budapest park, Tesla had briefly entered a state of heightened awareness which revealed how magnetic fields can interact. Even before he encountered Vedic science, Tesla had predicted, "Ere many generations pass, our machinery will be driven by a power obtainable at any point in the universe… Throughout space there is energy…it is a mere question of time when men will succeed in attaching their machinery to the very wheelworks of nature."[69]

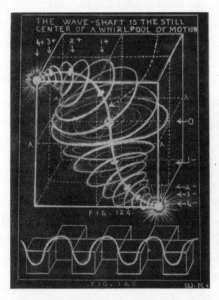

'The Wave Shaft' by Walter Russell
(with permission from The University of Science and Philosophy[70])

---

68   Grotz, Toby, "The Influence of Vedic Philosophy on Nikola Tesla's Understanding of Free Energy," published by Mountain Man Publications.

69   Ratzlaff, John, *Tesla Said*, Tesla Book Company, Greenville TX, 1984.

70   https://www.philosophy.org.

Toby Grotz' explorations ranged from studying remnants of ancient Anasazi culture in the American Southwest to reading the visionary Walter Russell's books.

Dr. Russell (1871-1963) once had a prolonged experience of heightened awareness. During his 39-day "cosmic illumination" he perceived how creation—matter, and the forces of nature—manifested via what he called the wave. Instead of a Big Bang, action is ongoing.

Dr. Russell saw the forces as two sets of dual clockwise and counterclockwise vortexes. One set of invisible vortexes is point-to-point, imposed on the other set base-to-base. Their action constantly integrates or disintegrates everything over time.[71] The wave of creation pumps space through endless cycles of compression and expansion, or gravitation and radiation.

When Grotz saw a schematic that viewed the "wave of creation" from the side, it appeared to him to be identical to patterns found in Navaho weavings. He wrote a paper "The Navaho and the Buddhist" speculating that ancient Indigenous peoples' symbols repeated the front view of motions of creation and disintegration Dr. Russell had described.[72] For instance, in Polynesian culture the image for creation is a set of clockwise and counter clockwise spirals carved in wood or stone.

Navaho rug. (courtesy of Toby Grotz)

"A Navaho rug symbol and (Russell's vision of) the wave explain the dual wave- and particle-like nature of atomic physics," Grotz said. "Viewed end on, the symbol becomes the yin-yang sign of eastern philosophy."

"It is also a pictorial representation of what Vivekananda taught Tesla—that space is filled with an ether, and prana acting on it created matter. It was not until eighty-two years later when Paramahamsa Tewari published his paper *The Substantial Space and Void Nature of Elementary Material Particles* in 1977 that the mathematical proof behind Tesla's statements and the descriptions in the Vedas was proven."

---

71   https://www.philosophy.org/scientific.html.

72   Grotz, Toby, *The Navaho and the Buddhist,* 1992.

Grotz speculated that ancient shamans who perceived the wave of creation could have used forces of nature to benefit their people. He took his ideas to physicist Thomas E. Bearden, who pointed out that sound vibrations could accomplish the same advanced technology that some inventors have done electrically and mechanically. Chanting could manipulate the wave of creation to cause physical effects, for instance. Sound accompanying a rain dance in the southwest desert could cause rain. Sound could heal the sick, or move large stone blocks by manipulating gravity.

Despite the esoteric studies, Grotz didn't neglect his engineering career. He worked in coal, natural gas, and nuclear power plants and in the aerospace industry.

Grotz worked with the late Robert Golka in1987 on "Project Tesla" at a mine near Leadville, Colorado. Using the same dimensions as Tesla's Colorado Springs tower, the two men built what was then the world's largest Tesla coil. They tried to resonate the Schumann Cavity—the spherical waveguide formed between the ionosphere and the earth—and recreate Tesla's system for sending

Paramahamsa Tewari of India and Stefan Marinov of Bulgaria at Hannover, Germany. (Jeane Manning photo)

power. Although their project fell short of its goal, both men appreciated Nikola Tesla's desire to make electric power globally available.

A 1987 conference in Hannover, Germany,[73] gave Grotz a chance to learn more about the dignified inventor from India who received an award there—electrical engineer Paramahamsa Tewari. Grotz had read Tewari's papers and corresponded with him.[74] Convinced that Tewari's mathematics and physics answered questions previously unanswered by quantum mechanics and the theory of relativity, Grotz began a professional relationship and friendship lasting for the rest of Tewari's life. When he arranged for Tewari to travel to an International Symposium on New Energy in Colorado, Jeane had

---

73   German Association for Gravity Field Energy, Hannover, Germany, 1987.

74   Raum & Zeit and Journal of Borderland Sciences articles.

opportunities to interview the inventor from India who had been honored in Hannover.

Tewari spoke about having been a young electrical engineer, sent to Canada to work for a nuclear power project,[75] and his more recent job of directing construction of a power station back home. (When he retired, he was executive nuclear director for the Nuclear Power Corporation in India's Department of Atomic Energy.)

Fundamentals of physics had intrigued Tewari from his school days onwards, but he took it further. He was a Sanskrit scholar, aware that he lived in a land of timeless wisdom, and his guru taught about a conscious ocean of energy from which material worlds emerge.

Tewari tested his hypothesis about energy by experimenting. In his spare time Tewari had built and tested[76] electricity generators to see if it's possible to tap into the universal energy surrounding us. He filed patents in India for his invention, the Space Power Generator.

His machine that won top prize in Hannover was similar to rotating "homopolar" generators being built at the time by Americans Bruce DePalma, PhD, Adam Trombly and Dr. Valone.

Tewari proved principles of his Space Vortex Theory, with mathematics:

- Space throughout the universe is an eternally existing, nonmaterial, continuous, isotropic (identical in all directions) fluid substratum. (Substratum means underlying layer or foundation.)

- When space moves in a circulating motion, it has a limiting flow-speed equal to the speed of light relative to the "absolute vacuum," and a limiting angular velocity. (Angular velocity is how fast it rotates.)

- The medium of universal space is eternal and endowed with motion.

Earlier scientists also held the concept of an ether—sometimes spelled as aether—eternally existing in space. Tewari's differs from the old material-ether

75   Douglas Point Nuclear Project, Canada.
76   Valone, Thomas, *The Homopolar Handbook*, Integrity Research Institute, Washington DC, 1994.

view. His theory describes how electrons are created, like swirls in a stream, out of a non-material background sea of fluid space.

Grotz became the driving force behind organizing an Institute for New Energy in the 1990s. One year a wealthy Colorado couple funded it, so the organization decided to send an engineer on a world tour. They chose Grotz because he had the know-how and instruments to test energy inventions. He visited notable inventors in the USA, Australia, New Zealand, Japan, Switzerland, Austria, England and India, to encourage them to bring their works to Denver, Colorado, for the symposium on New Energy.

Toby Grotz (left) organized symposia in Colorado; Stefan Marinov viewed measurements by test engineer Bryan Willson. (Jeane Manning photo)

Tewari's work most impressed Grotz. He returned to India repeatedly in the following years, tested prototypes of Tewari's magnetic generator and built a website for him.

Walter Russell's books had taught Toby Grotz that at least half of the universe's actions are hidden from our physical eyes, so he was prepared for Paramahamsa Tewari's writings about physics. The title of a book by Tewari sums up another of the two engineers' mutual interests: *Spiritual Foundations*.

Tewari's insights on fundamentals led to his invention. A newer model that he called the Reactionless Generator achieved a milestone validated by Grotz and other professionals. Their tests showed it producing more than twice as much electrical power than the operator put in to run it.

Tewari said "We have erred, though unknowingly, in our design of electrical generators and have remained in error for more than two centuries." Ever since British physicist Michael Faraday discovered electromagnetic induction, scientists have declared limits on how efficient motors and generators can be.

Efficiency "higher than unity" is ruled out (unity means the same amount of power comes out as anyone put into it).

If you tell an expert in standard electrical science that Paramahamsa Tewari's generator puts out more than twice as much as the carefully measured electrical input, you'll probably hear about Lenz' Law.

What if you design the machine differently...?

Russian physicist Heinrich Friedrich Emil Lenz (1804-1865) wrote about the reaction caused when an electromagnetic field is induced in a conductor such as copper wire. The direction of the induced field opposes the field that produces it. If you want excess power to come out of a standard motor/generator, forget it. By the way that motors and generators have always been designed, electromagnetic forces fight each other and defeat the machine's efficiency.

What if you design the machine differently? Tewari did. He created a novel configuration of the way the parts are put together, a different geometry.

That explanation is not enough to satisfy lecturers who remind you what the "laws" of physics say is impossible and add, "There's no such thing as a free lunch."

True, in a closed system. Think of a sealed box around the machine; energy can't leak in from outside the box. From the earliest steam engines to massive machinery today, the *law of conservation of energy* holds true wherever experts can measure exactly the amounts of fuel or electricity going in, and the heat or electricity or mechanical work coming out.

However, revolutionary motors and generators are being designed from a different mindset. Grotz and other engineers say the second law of thermodynamics must evolve to account for new discoveries such as data that Tewari obtained with his Reactionless Generator. The law should also take into account Tewari's Space Vortex Theory which describes the substance of space.

HOW MANY DOES IT TAKE TO EXPLAIN A LIGHT BULB?

Physicists cannot explain a certain phenomenon, Grotz tells us. If a light bulb is connected to a generator 186,000 miles away, one second passes before the bulb lights up. (That's the speed of light.) However, electrons are the supposed carrier of electrical current and energy, yet they only advance at a rate of micro-meters per second. In alternating current, first they move one direction on the positive side of the cycle, then back about the same distance during the negative side of the cycle.

"The electrons go nowhere... So how did the energy get into the light bulb? To answer the question, physicists have theories about 'group velocity of an electromagnetic wave' and what they call 'the Fermi speed of an electron' which invokes the theoretical physics of what they call Hamiltonian and Quantum Mechanics."

"Physics today does not understand electricity," Grotz concluded, "but then, they have not read the works of Paramahamsa Tewari."

Tewari developed a generator with efficiencies as high as 238%. The way he built the machine rendered Lenz's law partly ineffective, because the magnetic circuit in Tewari's design minimized back electromotive force and the resulting back torque. In other words, forces in his Reactionless Generator don't fight among themselves, so it has greater efficiency than 19th century laws say is possible.

Power comes in from surrounding space at every point in his circuit, so his generator doesn't operate within a closed system. The laws of thermodynamics are written for closed systems such as a steam engine's cycle.

OUT OF THE VOID

Austrian film maker Angela Summereder contacted Toby Grotz in 2017. She grew up beside a castle in Austria where inventor Karl Schappeller in the 1920s claimed he could develop a machine that would access free energy. Local farmers invested and lost their money. Summereder began a docudrama based on that failure, then she heard about Paramahamsa Tewari's success and asked Grotz to introduce her and accompany her film crew to India.

The resulting film shows the Reactionless Generator being tested, and visits the factory where an Indian corporation, Kirloskar, reproduced it and verified its operational characteristics. In the film, Grotz says Tewari's reactionless generator fulfils Tesla's prophecy "...mankind will attach his machinery to the very wheelworks of nature." The *Out of the Void* title comes from Tewari's conclusion that space is not empty and free power is real.[77]

Paramahamsa Tewari died in late 2017, but his writings about the physics of free power remain.[78] An obituary described him as "a life force that burned bright with innate curiosity about the physical and spiritual worlds." He was adventuresome and committed to the upliftment of society. As with many successful inventors, nature inspired him.

His son Anupam Tewari gave a talk about his father's work at the 2018 Energy Science and Technology Conference, in Idaho.[79]

Toby Grotz retired from industry but still investigates increased-efficiency electrical generators and transformers, as well as advocating for community gardens and other sustainable systems. He and his partner Kate relocated to a rural region to enjoy the small-town lifestyle that his email signoffs anticipated: "In the future, we will sit on our front porches with family, friends and neighbors, singing and playing acoustic music until the stars shine down upon us, undimmed by the fires of fossil fuels."

---

77   https://vimeo.com/ondemand/outofthevoid.

78   "Genesis of Free Power Generation," by Paramahamsa Tewari, B.Sc.Eng., Ch. 7 from *The Physics of Free Power Generation; it* also appeared in Explore Vol. 6, No. 3, 1995.

79   www.tewari.org.

CHAPTER 10
NOT YOUR ANCESTOR'S ETHER

We're revisiting the 'aether' in physics that was displaced by a materialistic worldview but may actually be the quintessence underlying physical reality.

—Beverly Rubik, PhD biophysicist.[80]

SCIENTISTS' VIEWS ABOUT A UNIVERSAL ENERGY FIELD vary widely, from "nothing to see, folks; move along" on one hand, to mainstream theories about a "Higgs field" or "zero-point energy" and to farther-out worldviews on the other hand. Some scientists question whether the energy embedded in the background is really electromagnetic, or if it's off the scale of the electromagnetic spectrum, or from a higher dimension. Others perceive a living but non-material field with an intelligence.

Technical experts in the new energy field startle us with their eloquence when they write about the universal source.

James F. (Jim) Murray III, for instance, is an inventor and consulting electrical engineer who deeply explored Nikola Tesla's concepts during four decades of research. Murray relocated from Oklahoma to further develop his own patented inventions in the laboratory of Canadian company Nonlinear

---

80   Rubik, Beverly, regarding USPA 2018 conference theme. Full quote: "We are also revisiting the 'aether' in physics that was displaced by a materialistic worldview but may actually be the quintessence underlying physical reality and the etheric body in radionics to explad the scientific perspective."

Dynamics Inc.[81] The NLDI company is inspired by Nikola Tesla's work, and aims to create novel hyper-efficient electrical technologies, focusing on electrical generators for micro-grids and electrical motors.

Nonlinear Dynamics Inc. team (photo courtesy of
CEO Jonathan Zerbin, in black shirt. Jim Murray is in plaid.)

A sampling of Murray's accomplishments:

- Re-discovered how Tesla magnified an electrical generating system's power.

- Discovered a power oscillation principle, which is distinct from current oscillation[82] and applied it to inventions such as his Dynaflux Alternator, an electro-mechanical power converter which operates on novel principles.

- Demonstrated fifty times more output than input in another of his inventions, the Switched Energy Resonant Power Supply (SERPS). In technical terms, when operating resistive loads, it had a Coefficient of Performance of more than 50.[83]

81   www.nldi.ca.

82   http://free-energy.ws/jim-murray.

83   Coefficient of performance or CoP is a measure of efficiency used to describe a heat pump, refrigerator or air conditioning system. It is a ratio of useful heating or cooling provided to work required.

Murray reminds other researchers that no limit has ever been placed upon the number of times that energy can be transformed from one form to another, nor upon the direction of energy flow associated with such transformations.[84]

THROUGH THE AGES, A GREAT POWER

Murray once published an essay about the ultimate power source. He wrote that the ancients believed a great vortexian power created the stars, the planets and all of the cosmos. That immense, swirling sea represented an infinite reservoir of raw energy and caused and created all things.

"Wilhelm Reich called this cosmic tide the orgone energy, and today some scientists refer to it vaguely as the zero-point potential."[85] Dr. Wilhelm Reich (1897-1957) was an Austrian-American scientist whose experiments revealed the reality of a life force energy that the ancients had called the ether and he called orgone.[86] It is present everywhere, in living beings and in Earth's atmosphere.[87]

Murray also quoted Tesla: "If the universe is dynamic, and we know that it is, it will not be long before man attaches his machinery directly to the workings of the cosmos and secures for himself an infinite and dependable source of power."[88]

Throughout history and pre-history, individuals have glimpsed the non-physical universal spiraling of energy, the invisible vortices that create matter

---

84 Nonlinear dynamics http://nldi.ca/our-story.

85 Murray, Jim, with permission, http://jimmurrayscience.com/jimmurray/page12.htm

86 A January 23, 2019 New York Times article "Mary Boyd Higgins, Wilhelm Reich's Devoted Trustee, Is Dead at 93," by Katharine Q. Seelye perpetuates the half-century of misinformation about Wilhelm Reich's legacy, positioning Dr. Reich as having been sex-obsessed which is untrue. His theories as a psychiatrist are a relatively small part of his discoveries, and he meant his orgone accumulator to enhance the life force for general healing purposes.

87 The life force concept has implications—the relationship between the life force and human health, and understanding the harmful effects of radioactive by-products of nuclear fission. Unnatural radiation hinders the flows of life force in our atmosphere.

88 Tesla, Nikola "The Problem of Increasing Human Energy," June 1900, Century Illustrated Magazine.

and destroy and create again endlessly. The same dynamic is everywhere, from the galactic down to sub-microscopic scale.

Visionaries imprinted the vortex onto human culture. They left clues ranging from the spirals of Celtic art to Maori carvings of clockwise and counterclockwise spirals.[89] Viktor Schauberger saw nature's universal tool—the vortex—in the growth of plants and in the flows of water and air. More recently, enlightened engineers are seeing how they can build generators that tap into vortex power found throughout the universe.

Although Nikola Tesla investigated many methods of doing that, "he left precious little information to guide us." Murray added that, despite the difficulty, it is the responsibility of humankind to forge ahead with noble intentions and discover the ultimate energy source.

The task will be enormous, Murray realized, "not because of the technical limitations, but rather because of our inability to think outside the existing paradigm and our fear of clashing with the vested interests who would gladly doom us all to a perpetual era of filthy, inefficient internal combustion engines."

Murray said humankind needs to progress beyond quantum mechanics to working with a fundamental and sublime power source. He suggested fellow scientists rely on the power of inspiration to aid them in the great undertaking.

"For if we desire to animate our industries with the Breath of the Universe, should we not first learn to listen to the subtle whisperings which it offers to every conscious mind? Before we engage in Celestial Mechanics, surely we must recognize the existence of the Cosmic Mind."

DIDN'T SCIENCE BURY THE ETHER?

For about a century, such ideas and inventions have gone unrecognized by the dominant worldview. However, over-unity inventions don't violate any law of nature if they tap into the background energy of the universe.

What is that universal energy?

---

89   Manning, Jeane "Did Ancient shamans know the secrets of the Wave?" Atlantis Rising magazine, October 2011.

Until modern times "aether" or alternately "ether" described an invisible immense, swirling cosmic sea of energy through which everything moves. Biophysicist Beverly Rubik PhD says the ether concept was "once a key scientific concept, then discarded and now re-discovered as 'neo-aether.' It deepens our understanding of the universe and expands science."[90]

Perhaps Albert Einstein's "spacetime" is the modern ether. In this book we use the spelling "ether" except when presenting the ideas of those who spell it as aether.

FAMOUS EXPERIMENT, BASED ON WRONG MODEL

In the late 1880s, American scientists Albert Michelson and Edward Morley looked for evidence of what their contemporaries expected to find, a motionless ether filling all space. They reasoned that if our journey around the sun passes through a fixed ether, as Earth moved through it there could be a detectable ether wind seeming to blow past Earth.

They used an instrument called an interferometer, that measures speed of beams of light by splitting and recombining them. If the beams interfere with each other, a screen would show patterns indicating that encountering an ether had delayed one beam.

The Michelson-Morley experiment failed to show the ether-wind effect they expected. However, the null result only showed lack of evidence for their mental picture of a non-moving ether. Their hypothesis was mistaken.

Twenty years later, Michelson won the Nobel prize in physics for work on optics, not for anything about the ether. Three years after winning the prize, Michelson said that Albert Einstein's relativity theory and an ether *can* be compatible.[91]

---

90   Rubik, Beverly, presentation for 2018 US Psychotronics Association.

91   Michelson, Albert, quoted in "Reuterdahl's commentary on Einstein…Minneapolis Morning Tribune, April 14, 1923, Michelson is quoted as saying that even if relativity is here to stay, we don't have to reject the ether. Researched by Aaron Murakami.

> ...closed the door to an ether, despite scientists who didn't agree.

After Einstein's theory gained popularity, however, science teachings closed the door to any concept of ether despite scientists who didn't agree. Over time, dissidents died off, while the mainstream viewpoint gradually elevated the importance of relativity theory and Michelson's and Morley's experiment as a reason to discard ether. Space was declared to be a vacuum and the name "vacuum of space" stuck.

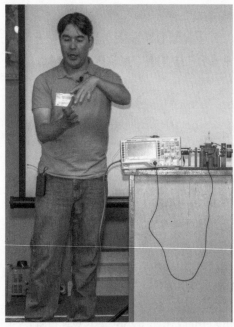

Aaron Murakami presents dissident views at his Energy Science & Technology conferences. (photo courtesy eMediaPress.com)

Nikola Tesla was one of the first dissidents to say that Einstein's concepts of empty space yet "curved spacetime" make no sense. "To say that in the presence of large bodies space becomes curved is equivalent to stating that something can act upon nothing," Tesla said. "I, for one, refuse to subscribe to such a view."

Many scientists today challenge the vacuum-of-space view, but peer reviewed journals won't publish heretical papers. At independent energy conferences, however, we do hear alternate theories such as in Aaron Murakami's presentation "Hacking the Aether."[92] They say that, even in Dr. Einstein's lifetime, certain experiments posed a threat to the belief that space is a useless vacuum. Dayton Miller, PhD, could not be easily

---

92   Murakami, Aaron, "Hacking the Aether: A Natural Philosophical View of the Aether, Mass, Space and Time." Presentation at 2017 Energy Science and Technology Conference, Hayden, Idaho. Speakers at that series, and at Steven Elswick's and Dr. Thomas Valone's conferences, present various theories often uniting electromagnetism and gravity.

dismissed, having earned a doctorate in astronomy at Princeton University and having a career teaching at the university where Michelson and Morley had taught.[93] Dr. Miller headed Princeton's physics department from 1893 until 1936. He was an author of textbooks and a member of the National Research Council in Washington, D.C. Dr. Miller performed thousands of careful experiments with an accurate interferometer and his results indicated an ether. More recently others have repeated Michelson and Morley's experiment using equipment of today and modern analytical methods.

Einstein had doubts about his own spacetime and empty-vacuum concepts. In a speech in Germany, he said that, according to his general relativity theory, "space without aether is unthinkable."[94] When he turned seventy, Dr. Einstein wrote to a friend, "You imagine that I look back on my life's work with calm satisfaction. But from nearby it looks quite different. There is not a single concept of which I am convinced that it will stand firm, and I feel uncertain whether I am in general on the right track."[95]

Our aim is not to pull Einstein's reputation down from its pedestal. After official science turned away from the ether, many technologies have been attributed to his concepts.

What could have been developed if official science had instead taken a path that incorporated non-material concepts such as the life force or ether? Maybe vitality, creativity and conscious connection to the Force—being in tune with divine realities—would be more valued. The natural world would not be commodified into numbers of board feet of timber to be harvested or gallons of water to be sold. Instead, nature might be respected for its intricacy, beauty and life-sustaining qualities.

---

93   Case School of Applied Science in Cleveland, Ohio.

94   Einstein, Albert, "Aether and the Theory of Relativity" Speech delivered May 5, 1920, University of Leyden, Germany, Einstein ended his speech with "Recapitulating, we may say that according to the General Theory of Relativity space is endowed with physical qualities; in this sense, therefore, there exists an Aether. According to the General Theory of Relativity space without Aether is unthinkable; for in such space there not only would be no propagation of light, but also no possibility of existence for standards of space and time (measuring-rods and clocks), nor therefore any space-time intervals in the physical sense. But this Aether may not be thought of as endowed with the quality characteristic of ponderable media, as consisting of parts which may be tracked through time. The idea of motion may not be applied to it."

95   Einstein, Albert, letter to Maurice Solovine, March 28, 1949 (quoted by Baneesh Hoffman and Helen Dukas in *Albert Einstein: Creator and Rebel* Plume 1972, p.328).

The idea of something energetic filling all space didn't remain completely banished for long, because quantum mechanics arrived during Dr. Einstein's lifetime. Quantum theory ushered in concepts such as zero-point energy, but physicists didn't walk across campus and tell the engineering departments. Students were not taught to think about engineering anything that is non-material.

Today, a growing number of scientists such as Dr. Elizabeth Rauscher network in informal groupings with names such as the "neo-aether movement."[96] They say the ancients had the wrong model, that's all. Whatever fills universal space is not unmoving like some ocean with no wave motion. Tesla's concept of ether and today's scientific model of a zero-point energy field are incredibly dynamic—more like the turbulent, foaming spray at the base of a high waterfall.

WHY CALL IT 'ZERO-POINT' ENERGY?

We wonder whether zero-point energy is the best moniker for the background energy. Quantum physicists call it that because they say its effects are detected as motion inside molecules even when matter is solidly-frozen at zero degrees Kelvin. Yet the phrase "zero-point" can sound negative and inadequate to describe an unlimited hidden energy that powers the universe.

Zero-point energy has already entered the vocabulary of millions of people[97] so it might prevail. However, properly naming a power source is crucial to how the public perceives it and how the science establishment treats it.

Science books and magazines give various definitions for the phrase zero-point energy, depending on their eagerness to dismiss the possibility of energy abundance while reinforcing the meme of scarcity. One popular magazine said it is energy produced by the miniscule movements of atoms at rest. Instead,

---

96   Elizabeth A. Rauscher is an American physicist, a former researcher with the Lawrence Berkeley National Laboratory, Lawrence Livermore National Laboratory, the Stanford Research Institute, and NASA.

97   Millions viewed the film *Unacknowledged*, or watched the *Thrive* movie, which used the phrase zero-point energy.

other scientists say the background sea of energy is the incredibly powerful *source* of those constant movements.

Our analogy is that nobody builds windmills to harness the tiny motions of tree branches or of grain stalks jiggling or swaying in the wind. Windmills instead tap into the movement of the wind itself. Likewise, revolutionary devices could tap into the universal energy itself.

MAINSTREAM SCIENCE LOOKS AT UNIVERSAL ENERGY FIELDS

Quantum theory describes electric and magnetic fields as they flow through space. At every frequency those fields or waves oscillate, and the motions never stop. Astrophysicist Bernard Haisch, PhD, describes the total result with an appealing phrase—a background sea of light.[98] However, he speaks conventionally when he adds the descriptor "electromagnetic zero-point field."

Dr. Rauscher prefers the term "vacuum state polarization." She worked at a national laboratory for many years and in research settings internationally. Her hundreds of scientific papers had been published in books and journals before she and professor Richard L. Amoroso, PhD, wrote *Orbiting the Moons of Pluto*.[99] Its complex mathematics goes over our heads, yet scientists open to exploring "complex higher dimensional spacetime" can appreciate its sweeping view of possibilities. The book also gave a theoretical foundation allowing for consciousness to act on the physical world, from the micro to the cosmic scale.

Even if we only catch glimpses of it, emerging science is awe-inspiring.

---

98   Haisch, Bernard, *The God Theory*. Red Wheel/Weiser, California, 2006, p 70.

99   Rauscher, Elizabeth, and Amoroso, Richard L., *Orbiting the Moons of Pluto: Complex Solutions to the Einstein, Maxwell, Schrödinger and Dirac Equations*, World Scientific Publishing, 2011.

CHAPTER 11
ROCKET SCIENTIST
SEES SUPERLIGHT

Superlight is the universal energy force in all nature.

—John Milewski, PhD.

JOHN V. MILEWSKI, PHD, IS A RETIRED LOS ALAMOS
National Laboratory scientist with a light-hearted approach to life. His grand-
daughter dubbed him "a wizard" and his family gifted him with a costume
to match, so on Halloween nights the scientist played the role as he greeted
trick-or-treaters at his door. The children in his Albuquerque, New Mexico,
neighborhood probably had no idea of the real-life identity of that wizard.

Dr. Milewski is a professional engineer, ceramics expert, inventor, and
former rocket scientist, now relocated to the East coast. He holds 30 patents,
has authored science books and has degrees from Notre Dame, Stevens, and
Rutgers universities. He earned a degree in chemical engineering, a Master's
in metallurgy, and a PhD in ceramic engineering.

Before his years at Los Alamos National Laboratory, he worked at Exxon's
research center and at a chemical company's rocket engine division. Later
he invented a process for growing single-crystal fibers of various materials.

He and his son Peter invented a revolutionary electric light bulb. Their
invention is displayed in the Smithsonian American History museum in an
exhibition honoring the six most significant ideas in the lighting field from
1950 to 2000. It was nevertheless rejected by the lighting industry. "Nobody

wants to make a light bulb that doesn't burn out *and* uses less energy," Dr. Milewski told Jeane, in tones of irony.

Joseph Peebles & John Milewski, PhD, partners in Open Technology LLC.

He is now working on a proprietary ultra-efficient energy transformer and refining his 'Theory of Everything.'

Dr. Milewski's theory predicts a new form of radiation that makes possible a future with clean energy magnetic lights, magnetic batteries and more. It fits in with the other dissident scientists' neo-aether movement. The visible light we're familiar with is part of the electromagnetic radiation spectrum. It has a strong electric component with a weaker magnetism part. The opposite of ordinary light is a *magnetoelectric* wave that Dr. Milewski named Superlight, in which the magnetic aspect of light is stronger than its electric part.

Superlight is the unseen force in nature all around us, Dr. Milewski said. We don't see or feel it because its frequency is so high and wavelength so short that scientists don't yet have laboratory equipment that can detect it. We do see its effects in the natural world. He explained gravity, inertia, so-called dark matter, the strong and weak nuclear forces and electromagnetism in terms of a magnetoelectric Superlight.

He figures that invisible light comes out of the extremely dense and very hot matter found at the centers of large galaxies. Superlight may be formed through energy exchanges from magnetic monopoles.[100] A monopole would

---

100   Dr. Milewski says the generating of Superlight is similar to how electromagnetic radiation is generated when electrons—electric monopoles—change energy state by dropping to a lower orbit. When magnetic monopoles change states by dropping to a lower orbit, they radiate a higher form of energy—magnetic light called Superlight.

be a north- or a south magnetic pole existing alone. Those monopoles are abundant in our galaxy, Dr. Milewski said.

HOW DOES THIS THEORY FIT WITH ZERO-POINT ENERGY?

Dr. Milewski described Superlight as a magnetic light that travels about ten billion times faster than regular light. Superlight is the sea of dynamic energy that fills all space and is the unified field that Einstein wanted to find.

"Superlight explains why gravity is a push (not an attraction force)," he explained in a paper.[101] "It is the same strong force that holds the nucleus of an atom together…Gravity and inertia are not intrinsic properties of matter but are a force artifact resulting from the presence of all-prevailing fields of Superlight." The density and geometry of matter change its manifestations.

DR. TILLER DID THE MATH

Theoretical physicists can't believe a new theory if mathematics doesn't support it, so we include a bit of history.

After Nikola Tesla invented radio (Guglielmo Marconi took credit), physicists looked for a mathematical model to explain radio waves. Nineteenth century mathematical physicist James Clerk Maxwell did solve an equation to explain forms of electromagnetic radiation such as light, radio, microwaves and X-rays. Maxwell had used regular (positive) numbers for that and found that when waves such as radio leave an antenna they radiate out into space in all directions to infinity at the speed of light.

---

101   Milewski, John V., *The T.O.E. that really covers everything. Superlight, a Dynamic Aether, Explains Pushing Gravity and Inertia, and Says No Neutrinos, Gluons or Dark Matter*, prepublication paper, Feb. 27, 2018

A second solution to the famous Maxwell's Equation uses negative numbers. For a century, algebra students were told to ignore what are called imaginary numbers because negative numbers have no meaning in the real world.

In the mid 1970's, a scientist at Stanford University, William A. Tiller, PhD, t found that the negative numbers solution to Maxwell's equation describes an opposite type of radiation; it comes from infinity and travels toward the point source from all directions and has a large magnetic and a small electrical part.[102]

Dr. John Milewski looked closely at Tiller's equations and concluded that this opposite type of light moves at the speed of ordinary light squared— ten billion times faster than the light we know and has a vastly higher energy density.

COULD IT SAVE THE MODERN WORLD?

Our society has made everything to run on electricity described as a flow of electrons. Dr. Milewski wondered what would happen if Earth's ionosphere gets whacked by some cataclysmic event and it becomes more difficult to access electrons. He speculated that electrical technologies might stop working.

As a backup, he suggested considering how to use magnetism as a way to generate *magnetricity* because it is a flow of magnetic monopoles, the magnetic analogue of electrons. Magnetricity may be a future energy source.

He credits entomologist Philip S. Callahan's experiments with tree roots for showing him that magnetic monopoles can be created. Dr. Callahan had insights about the way magnetricity is channeled from one place to another. It is the opposite of conducting electricity with copper wires.

---

102 Tiller, William A. (1975) "Positive and Negative Space Time Frames As Conjugate Systems" Proceedings of ARE Symposium AZ January

TREES AND ROCKS TAUGHT CALLAHAN

Philip Callahan, PhD, keenly observed the natural world. For instance, he discovered that paramagnetism is often missing from poorer soils around the globe. Paramagnetism is a property of materials that are attracted by magnetic fields.[103] His travels from the Amazon to the Far East taught him that the electromagnetic spectrum connects plants, insects, soil, and photons of electronic systems, to everything else.

Dr. Callahan also saw evidence of knowledge held by the ancients. He questioned why they built some megalithic stone structures with a para-magnetic granite exterior and diamagnetic wood inside. Were those choices of building materials deliberately used to accumulate magnetic force? His conclusions focused on the form of magnetism opposite to paramagnetic. Diamagnetic materials move away from magnetic fields. Paramagnetism is a stronger magnetic behavior seen only with certain materials like volcanic rocks, while diamagnetism is weaker and is found in all plant life and in our bodies.

Soil doesn't store magnetic fields. Dr. Callahan learned that instead it becomes tuned to *resonate with* magnetic frequencies from the sun and cosmos.[104] The takeaway for farmers and gardeners, he said, is to restore soil with loads of (paramagnetic) volcanic gravel plowed into dirt that is filled with organic materials.

WOODEN WIRING FOR MAGNETIC FLOW?

Dr. Callahan later discovered how magnetic monopoles flow within the roots of trees and upward. It's like a wiring system for magnetricity. Since water moves through the outer bark of roots and bark is a conductor of electricity, that makes it an insulator for magnetricity. Wood within the tree insulates electricity but conducts magnetricity.

---

103   Callahan, Philip S., *Paramagnetism* p. 35,78.
104   Callahan, Philip, *Nature's Silent Music*, Acres USA, Kansas City MO.1995, p. 195.

A society powered by electricity is familiar with seeing copper wire surrounded by an insulator. Dr. Milewski expects magnetricity to instead flow along something organic which would be an insulator for electricity but a conductor for a magnetic force. The inner material might look like plastic fishing wire. Copper on the outside would prevent magnetic forces from escaping in every direction.

Although our bodies are electrical systems, Dr. Milewski postulates that as Spirit we are magnetic. He saw that as a gift and speculated that our bodies and minds may need much more of that type of energy in coming times.

The tendency of magnetic fields to spread out through the universe explains instant communication on the spirit level, Dr. Milewski said, and there's no shield for emotions which are also magnetic energy.

Our sensitivities may be heightened in the coming years because of an influx of magnetic energy, Dr. Milewski speculated. Earth, along with all the bodies in our solar system, is in the early stage of moving through a highly energetic region of space. He referred to longer cycles, the sun's 25,000-year orbit of our solar system. Twice during each orbit, the solar system dips into areas of the galactic plane where our sun encounters different levels of magnetic energy and therefore gets more activated.

People are taught that our sun is a nuclear fusion reactor. Instead it is more like a magnetic light bulb, he said. When it is unusually energized, solar flares spit out energy toward Earth and impact the ionosphere that surrounds our planet. Earth's jet stream and weather are affected. As a result, weather is not normal. Dr. Milewski noted that emotions are also stirred up by those changes in the sea of energy Earth moves through.

To work with Superlight most efficiently Dr. Milewski looks to superconductors, which create strong magnetic fields. The semi-metal graphene might even be developed into a room-temperature superconductor. Graphene can be made in layers of an atom thick, much thinner than a sheet of paper. It's more electrically conductive than copper.

"By capturing Superlight, our preliminary calculations predict these energy-producing devices will generate a series of high-frequency, magnetic pulsed fields that will be specifically designed to send radiant magnetic power to whatever device needs it... for lighting, heating, cooling, cooking, washing or whatever..." Without pollution and at an extremely low cost.

CHAPTER 12
PEACE MAKERS IN THE
PARADIGM WARS

The wise use of potentially limitless energy
requires a shift of consciousness.

—Moray King, systems engineer

A PARADIGM IS A SET OF BELIEFS THAT FORM A PERSON'S view of the world. Some people hold on to their beliefs more lightly, while others closely identify with a given worldview.

Peter Graneau, PhD, researched physics at a prestigious American university for many years and had strong words about resistance to a new energy paradigm. "The indoctrination of physics students, their blind faith in what they have been taught by their elders, and the career punishment of those who challenge the consensus distributed in textbooks, is common knowledge since Galileo's time."[105]

American scientist Ken Shoulders (Chapter 19) noticed a common form of disbelief and named it the Goethe Effect. He cited German philosopher Johann Wolfgang von Goethe who said, "You only see what you know."[106] The effect can hinder any scientist who looks straight at a discovery and does

---

105 Peter Graneau PhD, "Is Dead Matter Aware of Its Environment?" Frontier Perspectives magazine, Vol 7 No. 1, p. 51.
106 Shoulders, Ken, "Scientific or Technical Belief," essay emailed to Jeane Manning, 2007.

not recognize it because it would disrupt the teachings in currently accepted fields of science or engineering.

Shoulders said the expert who does recognize a disruptive finding often ends up working with it in isolation, self-funded and forced into a long silence about it.

His discovery—microscopic clusters of electric charge—was an enigma that defied presently accepted rules of science. He said he found it nearly impossible to communicate such radically new findings to many of his peers. The more they had learned about their specialty, the less they seem able to accept new concepts near their field.

When Shoulders told trained physicists about charge clusters' anomalous behavior, they cut off communication "as if a defense mechanism of the older, more familiar philosophy comes into play." In contrast, when others with less training heard Shoulders express a new idea, they began work on novel devices based on the information. He witnessed these open-minded people mentally conjure up new energy devices just from learning, for instance, that an effect called mass modulation causes electrons to launch into a form of self-propulsion.

'USE THEIR ACCEPTED LANGUAGE'

Systems engineer Moray King of Utah investigates inventions reported to tap into a new energy source and looks for common threads to explain them. He wants more scientists to do experiments that will prove, disprove, or at least shed light on the emerging knowledge.

Attempting to convince his professional colleagues in industry to investigate what they have been taught to shun, he uses their accepted language. Instead of speaking about an ether, he has for decades been digging up peer-reviewed science papers about zero-point energy.

Meanwhile, he is dismayed to watch the quantum gravity people fighting word wars with the string-theory people. The source of anger seems to be "if you're right then I'm wrong, and I can't be wrong; I've got to know it all."

King has a strategy for dealing with critics of a disruptive development. He asks the critic which paradigm they personally invest in. Calmly recognizing different personal paradigms is more productive than arguments.

He outlines some beliefs held by scientists and engineers:[107]

Classical physics is the standard paradigm for engineers. Its beliefs include a fixed background in space and time, one fixed future and all systems evolving like a perfect machine.

Einstein's **Relativity** theory was the next paradigm in the 20th century. As you saw in Chapter 10, it dismissed an ether because of a flawed experiment.

Quantum mechanics brought in items such as zero-point fluctuations and "non-local entanglement" that, King quips, would have Einstein rolling in his grave.

Quantum gravity vs. string/Brane theory has been a hot debate.

Consciousness transformation is where King hopes it's all leading.

"We're in the midst of a paradigm war right now," says King. "What's starting to emerge from that is the true existence of a fourth dimension of space. Further out are new paradigms of what is consciousness."

What happens if a scientist starts to believe in an advanced worldview such as consciousness transformation? Typically, somebody from an earlier paradigm such as classical physics or relativity dismisses that scientist as a kook. The so-called kook, on the other hand, looks back at those holding the older viewpoints and sees them as outdated.

Since most engineers are in the standard, classical worldview they typically don't believe zero-point energy is a reality. Even scientists who do recognize the term zero-point energy disagree about what it is. And when they could possibly harness something for clean energy, definitions matter.

King listed differing views about zero-point energy:

- It doesn't really exist; it's just virtual, theoretical.

- It propagates throughout space like heat.

- It exists only in three-dimensional space and arises from nowhere. "When you're saying that, you're violating (the physics law of) conservation of energy," King points out.

---

107   King, Moray. on Gary Hendershot's podcast segment *Paradigm Shifts.*

- "Or you can choose to interpret that it enters from an actual physically real higher dimension of space. This is where the paradigm of zero-point energy is leading us, as we apply more advanced physics to what is going on..."

Why do respected physicists get emotional over whether to allow zero-point fluctuations from a 'higher dimension' to enter their quantum gravity theories?

'IT'S ABOUT THE MATH!'

There's a practical reason, King explains. They are tied to using "computational machinery" that takes many years to learn. It's about the math! Bringing in those zero-point fluctuations makes it impossible to do the calculations that physicists are trained to do. Theorists who describe fluctuations popping in and out of a vacuum of space would need a mathematical model of where those fluctuations come from.

Quantum mechanics had removed the vacuum fluctuations by a trick called "renormalization." The mathematical trick was a way to eradicate troublesome "higher-order terms that would be infinities" in mathematics. If the vacuum fluctuations were allowed, King points out, physics would have to start over.

"This is the primary objection why we can't allow the vacuum fluctuations to enter the system. It would cause a paradigm shift."

PARADIGM WARS GET VICIOUS

Moray King notices that when the results of someone's experiment violates accepted laws of physics, the community that believed in the old paradigm throws out the scientific method. "Instead, scientists act like people, human beings. The new claim is ignored, ridiculed and rejected. They will not change their minds."

Science historian Thomas Kuhn, PhD, showed in his book, *The Structure of Scientific Revolutions* that eventually the resistance dies out and a new generation takes over.[108] In the meantime, words like "fraud" are thrown around and can ruin someone's career.

In order to encourage clearer communication between antagonists, King categorized types of so-called fraud:

- Type 1. An intentional scam.

- Type 2. Mistaken measurements.

- Type 3. A true energetic anomaly was measured, but (a) the person has poor business practices, a problem common with an inventor who lacks funds to complete the invention. Or (b) the explanation could be incorrect. That also is common, King says, when nobody understands where the energy anomaly comes from.

Moray King's key messages for online chat rooms are:

> When you call someone a fraud, name the type of fraud you think they're perpetrating.

Understand the paradigm you're in, so when you say "disobeyed the laws of physics" you know which paradigm those laws are in. "That will...allow us to be open-minded to each other's ideas because we're not busy defending ourselves..."

King's final message is "build the energy machine that changes the world."

> **"With optimism, hope and openness, great things can happen."**

Peaceful paradigm change can start with individuals.

Farlie Paynter is a Canadian researcher who has spent his own money to help inventors on several continents for at least twenty years and has seen situations that arise with both inventors and investors. Attitudes may have created many of the obstacles faced, he says. "But with optimism, hope and openness, great things can happen."

---

108   Kuhn, Thomas S., *The Structure of Scientific Revolutions,* University of Chicago Press; 4th edition 2012.

"The only solution is (based on) how we treat each other. We have to send love toward big oil or we can't move forward. We'll have to work with them, help them change over to non-polluting energy sources."[109]

---

109   Paynter, Farlie, conversation with Jeane Manning, Jan. 22, 2017.

CHAPTER 13
MAKING SENSE OF
THE UNIVERSE

*The universe is far more magical and beneficent than our
physics professors, government spokespersons, and media
conglomerates have led us to believe.*

—Paul LaViolette, PhD.

STARGAZERS THROUGHOUT THE EONS HAVE WONDERED
how the universe began. Paul LaViolette, PhD, has an explanation that sheds
light on how new energy inventions work.

Dr. LaViolette is an American systems scientist who was living in Athens,
Greece, when we most recently interviewed him.[110] His approach to micro-
physics accounts for forces in a unified manner and resolves long-standing
problems in physics.[111] His theory leads to a new understanding of force,

---

110   Interview by Jeane Manning with Paul LaViolette, March 2018

111   For instance, the subquantum kinetics cosmology led Dr. Paul LaViolette to make
successful predictions about galaxy evolution that were later verified with the Hubble Space
Telescope. He is credited with the discovery of the planetary-stellar mass-luminosity relation
which demonstrates that the sun, planets, stars, and supernova explosions are powered by
spontaneous energy creation through photon blue-shifting. With this relation, he successfully
predicted the mass-luminosity ratio of the first brown dwarf to be discovered. More recently,
his maser signal blue-shifting prediction has found confirmation following publication of the
discovery of a blueshift in the Pioneer 10 spacecraft tracking data.

acceleration, and motion; the energy potential field (ether concentration gradient) is regarded as the prime cause of motion.[112]

Paul La Violette, PhD, systems engineer & author (photo by Juha Hartikka, Finland)

The universe seems friendlier in light of his theory. "You don't have black holes that can swallow you, you don't have a 'big bang.' It's like the beauty of nature, a continuation of growth and the way things happen..." He views the creation of the universe as a sacred process of ongoing evolution.[113]

Dr. LaViolette's theory also pictures how devices could put out more energy than any measurable energy put into them.[114]

His life story shows what makes someone leap beyond what they have been taught and speak their different truth regardless of cost. Dr. LaViolette's parents were both scientists, his mother a chemist and his father a physicist and electrical engineer. Before Paul was born, they worked on the Manhattan Project in Washington state and afterward his father worked for a General Electric atomic power laboratory.[115]

Paul absorbed dinner table discussions of his father's engineering tasks. At an age when children draw cats and bunnies, eight-year old Paul doodled nuclear reactor systems.[116] He was ten years old when his father took him to Chicago for an Atom Fair. While his father attended lectures, Paul wandered

---

112 Dr. LaViolette's presentation for Steven Elswick's 2019 ExtraOrdinary Technology Conference.

113 LaViolette, Paul, *Parthenogenesis, A Sacred Journey,* video of nspiring music and views of a re-creating universe, can be purchased at etheric.com.

114 LaViolette, Paul, *Subquantum Kinetics: A Systems Approach to Physics and Cosmology,* Starlane Publications, Niskayuna, NY, 2013.

115   Fred LaViolette worked at General Electric's Knolls Atomic Power Laboratory.

116   LaViolette, Paul, "Tracing the Origins of Subquantum Kinetics," 2008, see http://www. etheric.com.

through exhibits ranging from rare earth elements to fuel rods. Newspaper reporters looking for a story photographed the precocious child.

As a result of the mentorship, Paul LaViolette thought in terms of processes from an early age. When his third-grade class visited a milk bottling plant, an automated conveyor belt fascinated him and he afterward invented and sketched assembly lines where a product would change and modify.

When given a chemistry set, he began ordering from a chemical supply house and also launched into pyrotechnical experiments and rocketry. He credits his father for teaching how to think abstractly. They shared a desire to explore the unknown and curiosity about nature.

Dr. LaViolette's mother taught him to think independently, unafraid of being different. "Following her example, I acquired the courage to fight, to stick to my ideas even if they were challenging mainstream thinking."

> "Following her example, I acquired the courage to fight, to stick to my ideas even if they were challenging mainstream thinking."

Conversation during his aerospace engineer uncle's visits further stimulated Paul's creativity. His uncle did rocket design, researched ultrasound, and developed a type of accelerometer. A later version of it measured the moon's gravity.

Being in a future oriented family is not enough to propel someone outside the box of conventional concepts such as everything is just based on structures. A mystical experience jolted Paul's worldview and showed him other ways of knowing.

While studying for his bachelor's degree in physics at Johns Hopkins University, he experienced an altered state of awareness and unexpectedly felt that higher intelligences were conveying information about existence— nature at its most fundamental level is in a state of flux and what we call structures are simply steady-state patterns in that flow.[117]

The experience started him on a quest to develop a theory of natural systems that included living cells, bodies, solar systems and beyond, on the tiniest

---

117 Mystical experience is described in the prologue to Paul LaViolette's book *Genesis of the Cosmos.*

scale and in the macrocosm. Eventually he would conclude that all those entities persist as stable structures because they are continually recreated from an underlying flux of events such as the orbits of something smaller.

He later discovered that a set of simple concepts could explain natural phenomena on many scales. Systems were nested within systems within systems in a vast hierarchy.[118] He made an analogy. "Everything is the same; it's just played in a different musical key."

BIRTH OF A THEORY OF EXISTENCE

The physics epiphany didn't come at the time of the mystical experience. He first completed his physics undergraduate work and decided to get a master's degree in business administration from the University of Chicago. During a course in organizational psychology, he discovered general system theory, an academic subject that had not yet caused any stir in physics.

Other threads of learning appeared:

- A paper by Albert Einstein said Einstein believed particles were *not* the points in space being described by other scientists but were diffuse structures called "bunched fields."

- Articles in science journals discussed a certain reaction that can create chemical wave patterns.[119] It was part of scientist Ilya Prigogine's work proving that certain systems classified as open chemical *reaction-diffusion* systems can create concentration patterns named *dissipative structures*.

In academia at that time the idea of a chemical wave was new. Previously, only mechanical waves were discussed. Dr. LaViolette wanted to take the newly circulating ideas and apply them to physics.

Chemical wave processes occur between molecules. To explain how something at the material *quantum* level—subatomic particles—could form, he

---

118   LaViolette, Paul, "Tracing the Origins of Subquantum Kinetics," p. 4, 2008.
119   Belousov-Zhabotinskii reaction.

would bring similar ideas down to the *subquantum* level and find a different underlying layer. That would fill all space yet be invisible to instruments that detect everything else.

Paul LaViolette's "aha" experience was his realization that subatomic particles could also be "dissipative structures, concentration patterns forming in an underlying medium that engaged in reaction-diffusion processes." Such structures could be the bunched fields Einstein wrote about.

Dr. LaViolette was able to understand how the structures emerged and maintained themselves. Subatomic particles would form as distinct structures in much the same way that chemical waves would form in the special reaction he had learned about.

The ideas were foreign to anything he had been taught by his professors. No one thought subatomic particles needed any underlying flux or process to maintain their structure.

It was 3:00 AM in Chicago when Paul LaViolette realized he had an entirely new development of theoretical physics; he could see how creation can be perpetually happening. In his memorial tribute to his father he described the excitement of that moment. And its humorous aspect.

He had to phone his favorite mentor regardless of the late hour. His father woke in a hurry when he heard Paul's excited, "If I don't live until morning, you should know about this!" The son had to assure his father that he hadn't been in an accident. "It's just that I may have made a great discovery in physics."

> "If I don't live until morning, you should know about this!"

The thrill of discovery overturned Paul's sleep schedule for days and he worked through the nights writing about it. In a sleep-deprived altered state, he was able to sense what he describes as nature's kundalini everywhere. "Walking across campus, I could sense this etheric flux in all things, in the trees, in the rocks; everything seemed to be patterns formed in this vibrant flux."

Going from his "aha" moment to having a theory published in respected science journals took decades of work. For a start, he took kinetic equations used by chemists to represent chemical reactions and reduced those equations to only three lines of mathematics representing etheric processes.

"It has an elegance, a sort of beauty about it; it's so simple. And from this one set of equations that has certain assumptions with it so that you understand what it's about, you get so many predictions!"

He envisions the processes underlying the physical world as "a transmuting *ether* consisting of interacting etherons." Those particle-like non-material entities exist at a subquantum level more fundamental than the level where quantum physics stops. Etherons are involved in creating subatomic particles from the ether's reaction-and-diffusion processes.

Don't call that energy, Dr. LaViolette warned. At the subquantum level there are processes and activity. What is usually meant by energy begins at the quantum level where the physical level starts and things can be detected. To portray the creation of the universe, he produced a video *Parthenogenesis: A Sacred Journey*.

"For all we know, etherons could be conscious entities just like us. Whatever they are, all we know is there are types (which he named with letters of the alphabet) and they react according to certain patterns… But they could have personality; they could have their own lives. It's not going to affect anything at our level. One thing we know is that they're born and they die and they react and interact and that's what generates our universe."

"What physicists view as empty could be full of consciousness."

"What physicists view as empty could be full of consciousness."

At first, Dr. LaViolette called his approach an alchemical ether theory. He later gave it the name *subquantum kinetics* which sounds more acceptable to physicists.

Biologists may grasp his ideas more easily. Lay people can also understand that our bodies are sustained by ongoing chemical reactions and chemical diffusion, or transport as in the circulatory system.

What did his favorite mentor, his father, think about Paul's disruptive theory?

Dr. LaViolette admits that at first his father was hesitant to accept the subquantum kinetics approach. He had always dealt with conventional physics concepts. Whenever Paul would visit, his father tested the theory by asking how it would solve problems that mainstream theories address.

"Every time, I came back with a satisfactory answer or with evidence showing that my theory offered a superior, more plausible explanation. With his background in nuclear engineering and chemistry, he readily grasped the concepts, probably more quickly than most physicists whose early training taught them to think mainly in mechanical terms."[120]

As Paul honed the theory, his father became his sounding board and entered the wondrous realm of subquantum physics. Eventually Paul noticed that his father's eyes lit up when the two of them explored its implications.

A theory is judged by whether it predicted something that was unknown at the time the prediction was made. When we last spoke with Dr. LaViolette, he said subquantum kinetics had thirteen of its published predictions verified either through data coming in from astronomers and other scientists that matches his earlier prediction or by results of experiments.[121]

ENERGY AMPLIFICATION ALLOWED

He coined another term in addition to etherons. *Genic* as in generative means that the ether generates something new and photon energy increases when certain conditions allow. Energy in an open system can either increase or decrease. Genic energy is the over-unity amount of increased energy.

Dr. LaViolette uses an analogy of a bank account gathering interest, to portray excess energy. The amount of money deposited is like the amount of heat energy inside a star. If you could count all those photons and go back for a recount much later, the percentage of increase would be like interest in a bank account.

---

120   LaViolette, Paul, "Tracing the Origins of Subquantum Kinetics," 2008, see http://www.etheric.com.

121   Dr. LaViolette's predictions are summarized in the back of his book *Subquantum Kinetics*, and on http://etheric.com/predictions-part-ii-physics-and-astronomy.

As in nature, the excess energy from universal energy technologies "may ultimately be traced to the Prime Mover that animates the universe…the vast reservoir of ever-present transmutive etheric activity that fills all space," Dr. LaViolette writes.

> **"Inventors of over-unity energy devices have learned… what nature does all the time…"**

That can be why successful over-unity devices require no fuel to produce their power output. Instead, extra power comes from "energetic processes."

"Inventors of over-unity energy devices, it seems, have learned a way to generate genic energy, what nature does all the time," he writes. "By astutely observing nature and following their intuition they have succeeded in far surpassing nature in terms of the rate of genic energy generation."

Dr. LaViolette's visuals for his presentation at the 2017 Energy Science and Technology Conference contained an image of a patchwork quilt, representing physicists' attempts to unify multiple physics theories conceived independently.

Exiting the closed universe paradigm will be a big jump for physicists, even with a new approach such as subquantum kinetics welcoming them. However, when enough scientists make that leap and start over in an open-system paradigm, perhaps humans could travel to the stars.

PART III
IT'S RAINING
INNOVATIONS

CHAPTER 14
SOLID YET IN MOTION

Some of the biggest 'problems' in physics—like identifying
dark matter—are Creation's radical abundance in disguise,
misunderstood by a species accustomed to lack and firewood.

Graham Gunderson

ELECTRONICS ENGINEER GRAHAM GUNDERSON PREDICTS
humankind is on the cusp of an energy revolution. He is in a position to
know, since he specializes in advanced power analysis methods for testing
claims of energy output as well as invents his own energy devices.

More than twenty years ago, his role in
the field had a challenging start—rebuilding
a novel electrical generator in a barn. His
aunt knew that her nephew Graham was
unusually skilled in assembling electronics
and mechanical things, so she connected him
with a man who had a problem but not the
mechanical skills to solve it.

The man lived on a cattle ranch in southern
Oregon and had previously hired inventor
Barbara Hickox to build a generator. The
rancher had believed Hickox would be able
to build a generator that put out so much
electricity it could power itself and more.

Barbara Hickox, inventor.
(photo by Jeane Manning)

Hickox had a U.S. patent on an electrical dynamo she had made with unusual mechanical features.[122] As a girl she had watched her father fix cars and discovered her own love of tinkering with mechanical devices in a garage. She was later a protégé of the late Howard Hughes (1905–1976) the eccentric inventor who started Hughes Aircraft, she told Jeane. The photo above was taken after a meeting in the American southwest; Barbara Hickox was showing a prototype of her invention to a few engineers.

Later in Oregon, she and the rancher had a dispute and he fired her. He was left with only a box of parts to show for his investment, because her patent wasn't a set of instructions. When Graham Gunderson arrived, he used the limited tools in the barn to build a version. Its efficiency was less than impressive. "That's not to say it was representative of Barbara's invention. It was just my own version," Gunderson told interviewer Aaron Murakami.[123]

Since he couldn't take the project further, Gunderson retired from that job to make his living by repairing audio equipment. However, the ranch experience had piqued his interest. Over the years his breakthrough energy research turned full-time; experts recognized the quality of his thinking and his professional skills. Later in this chapter we look at demonstrations he gave at conferences in Idaho, and his new-paradigm attitude.

SOLID OUTCOMES

We asked Gunderson for an update on the branch of breakthrough energy research called solid-state. He began by predicting outcomes of an energy revolution:

- Electric cars will recharge themselves while they drive.

- Personal electronics will stay on forever without charging.

- Coal burning will be seen as ridiculous.

---

122   Hickox, Barbara, Electrical dynamo United States Patent 4249096.
123   www.energyscienceconference.com.

- Alternatives should be cheap when electricity is made anytime and where needed.

"Even hydro dams can't compete. In some cases, rivers could regain their native salmon. And the nasty fracking, flaring and poisonous oil spills start to become a thing for history books, not breaking news. It truly is a revolution."

Energy is everywhere because motion is everywhere Energy is everywhere because motion is everywhere, he said. "When you're at absolute zero (on the Kelvin scale) that's really when everything truly stops, truly freezes. Well, an ice cube is 273 degrees hotter than 'absolute stop'…very hot, on the absolute scale of temperature."

His ever-present sense of humor slips in: "Physical stuff like air, ice cubes and argyle socks are made of molecules, and all the molecules are always moving, shaking around, like a permanent earthquake."

Temperature is our measurement of the strength of that molecular movement. The hotter something is, the more its particles move. "It's energy we can use instead of deadly coal, instead of dams that kill fish, instead of expensive wind turbines…Anything at room temperature is full of this free motion; it's ready to be captured."

The last piece of the puzzle is how to capture the energy and use it. Technological advances have made it possible.

Gunderson reminds us that humankind couldn't access practical amounts of oil before drilling equipment was developed. Oil just sat in the ground. "Then someone started drilling and boom, we have gas stations and airliners. Now the same thing is happening with ambient energy and this time there's no exhaust." (Ambient means the energy comes from the surrounding space.)

Jeane Manning presenting categories at Energy Science & Technology conference 2018 (Chester Ptasinski photo)

Some inventions in the category called solid-state tap into ambient energy. A solid-state device is simply one that doesn't appear to have moving parts.

There's no rotor or wheel going around, no gears, pulleys or belts and no pistons, fan or bearings to wear out.

"It just looks like an electronic circuit board with some stuff on it, somewhat like the inside of a TV set," Gunderson explained. Motion is energy we can use instead of oil, solar or batteries. It's always there, nothing to buy, no waiting for sunshine, no clouds in the way.

All we have to do is get that random motion going in an organized direction and it becomes a force pushing in that direction. If we use that force to push electricity around, then we've created a battery that never ever needs to be charged. It's permanent power. You can't call it a battery anymore, because it never dies."

The random movement of microscopic particles in a liquid or gas caused by collisions with surrounding molecules is *Brownian motion*.[124] University scientists from the United Kingdom and Switzerland in October, 2017, found a way to turn that ubiquitous motion into a force that pushes electricity—a material using ambient energy to make miniscule magnetic movements that generate electricity.[125]

Gunderson says their discovery is similar to what American inventor Floyd Sweet created in the 1980s. Sweet's fuel-less machine put out more than 500 watts of free energy around the clock. As a result of taking heat from its surroundings to run it, Sweet's device became cold as it sat on a desk pumping out electricity. And no one could explain it, Gunderson said.

Research into tapping motion from the surroundings to power nanoscale (extremely tiny) devices is becoming mainstream.[126] Gunderson notices, however, that emerging research in the mainstream community is directed more at increasing computing speed than replacing sources of pollution. "Faster computers aren't much if you're still burning coal."

---

124   Graham Gunderson notes that "You can see this energy with your own eyes, by logging onto YouTube and searching 'Brownian Motion—Nanoparticles in Water.'"
125   Scientists from University of Glasgow and University of Exeter along with ETH Zurich (the Swiss Federal Institute of Technology) and the Paul Scherrer Institute, also in Switzerland.
126   https://phys.org/news/2017-10-scientists-ambient-motion-nanoscale-devices.html.
Sebastian Gliga et al." Emergent dynamic chirality in a thermally driven artificial spin ratchet," *Nature Materials* (2017

> "Charging the latest iPhone with electricity from a 50-year old coal plant is…insane."

The technical team he works with finds ways to do cutting-edge research while using materials common on the open market. "We are leveraging today's incredible technology with cheap, affordable FPGAs to control energy and get these systems working in ways that weren't possible before." (FPGAs are Field Programmable Gate Arrays—reprogrammable semiconductor devices.)

"Because when you think about it, charging the latest iPhone with electricity that came from a fifty-year old coal plant is really kind of insane."

"One thing that's more insane: *investors* haven't found out about this yet. This is the technology that will obsolete oil and it's not generally on the radar in the investment world…"

MINDFUL VALUES TRANSCEND MONEY

At the 2015 Energy Science & Technology Conference in northern Idaho, Gunderson admitted to having considered taking the working-in-secrecy route. His conscience soon ruled that out.

He had made a technical breakthrough with an experiment—toppling the prevalent belief that it is impossible to take a random, scattered energy and sort it into something that isn't scattered. Although his experiment only showed a small energy gain it demonstrated that technicians *can* use ambient random noise or heat or perhaps zero-point energy and rectify that (line it up into a form that's polarized and able to do work.)

Gunderson had been tempted to remain close-mouthed about certain details of his discovery. After all, he had spent his money and many exhausting nights on that experiment after his full-time day job. Why should he show up at the conference and give away something that came out of his own mind? Why risk exposing his brainchild to the world?

Numerous inventors have felt that urge to keep their "baby" close to their chest. "You made it. You want don't want someone to steal it."

However, Gunderson knew that a grasping mentality, inventors clinging to secrets, had kept energy technology breakthroughs in the dark for too many generations. His goal on the other hand was to give other "builders" enough information to be able to replicate the experiment. So, he shared details openly.

At one conference Gunderson showed how even a plain, small transformer can be made to give a measurable energy gain, meaning more total measurable work than he pays for on the input. A transformer is a non-moving device which transforms electrical energy from one circuit to another without a direct electrical connection.

His transformer was configured to create a moving magnetic field—a motional field that didn't stay in the same place. That slight motion was enough to cause electrical over-unity results. He remarked that it seemed to come alive and breathe.

A larger device that Gunderson called his Magnetic Implosion Transformer was one of several demonstrations of more output than input the next year at the annual conference in Idaho. On a large screen, the audience's technicians saw watt meters showing the 1.53 watts input and 9.43 watts output. Meanwhile, oscilloscopes and wave form analyzers added more technical information to prove the magnetic field was moving, not just switching on and off, although no magnet was itself moving.

The device's output lit a hot incandescent bulb and was 570 per cent more than could be accounted for in electricity measured as input to the device. He could even dial the input power down to zero watts and the light bulb remained lit.

"This transformer design is in its infancy, so who knows where you might be able to take it," he said. Avoiding hyperbole, Gunderson described it as an initial demonstration. Audience members called it a very exciting science experiment.

He cautioned, "Keep in mind that the results are measurements and I'm still in the process of validating the apparent over unity. By using two distinct methods of measurement to compare the output work to the input that I pay for, everything is showing me the same numbers, even with two different kinds of power analyzers."

"Although these results have been consistent and it is an astounding demonstration of over unity, one of the highest ever publicly demonstrated, I'm not quite sure why it works this well," he admitted.[127]

Gunderson said motional magnetic fields, involving a magnet whose field can move even though the magnet stays still, are the common denominator in some legitimate technologies that demonstrate over-unity performance.

> Whenever a magnetic field is moving, electricity is generated. Ordinary generators require propelling magnets by powering an engine with a fuel. In Graham Gunderson's Magnetic Implosion Transformer, only the magnet's field moves while the magnet holds still. Part of the force propelling its field appears to come from the ambient energy that surrounds everything. "If you adapt this to various coils," he said, "this ability to use ambient energy to fuel the motion of the magnetic field can and will produce electricity."

Gunderson spent countless hours studying claims and diagrams of inventor Floyd Sweet's solid-state device. Results of experiments on Gunderson's own workbench indicate some truth to Sweet's claims of having conditioned barium-ferrite magnets to self-resonate when hit with specific frequencies.[128] Sweet's claims of million-to-one output to input ratios were unusually high, but Gunderson at least showed that there are little-known properties to those magnets. Their magnetic fields move because of built-in instability within barium atoms.[129]

The next chapter reveals renewed hope for devices such as Sweet's.

---

127   http://emediapress.com/grahamgunderson/mit.
128   http://grahamgunderson.com/ou.
129   Graham Gunderson presentation to 2016 Energy Science and Technology Conference, Hayden, Idaho.

CHAPTER 15
MANELAS' ELECTRIC CAR

*New physics shows that the quantum energy of space—
zero-point energy—is the power that leads to
and sustains magnets.*

—German physicist Thorsten Ludwig, PhD.[130]

THREE SENTENCES IN A DECEMBER 6, 2014 NEWSPAPER
indicate a surprising achievement. The words published in the Lowell,
Massachusetts, newspaper were written by electrical engineer John Manelas
within a longer tribute to his father, Arthur.[131]

"The past 15 years he dedicated his life to his lifelong dream of supplying
free energy to the world. He developed a system that produced more energy
than it consumed. A true over-unity energy device ..."

Arthur Manelas' accomplishments were known to only a select group
during his lifetime. Keeping a low profile allowed him a normal life, which
included restoring muscle cars and enjoying his family. Colleagues who had
respected his privacy praised him publicly after his death, describing a "quiet
achiever" and "one of the most brilliant and likable people I have ever met...
smart, private and above all willing to explore in areas others fear to tread...a
generous spirit and all about helping humanity."[132]

---

130   Ludwig Thorsten, Abstract for presentation at 2019 ExtraOrdinary Technology
Conference.
131   Arthur Manelas obituary published in Lowell Sun, Dec. 7, 2014.
132   Mark Dansie, Revolution-Green.com.

Manelas had immigrated from Greece, earned a degree in electrical and mechanical engineering, and worked in the U.S. space program during the Apollo 13 mission. We'll pause his story there while revealing the backstory to the invention that powered the batteries in Manelas' electric car, because his creativity had been sparked by reading about Nikola Tesla and the invention of the late Floyd Sweet.

FLOYD "SPARKY" SWEET

Controversy rocked the meeting room, the first time Jeane heard about Sweet's invention. The room was a blue-carpeted and chandeliered auditorium in a large Boston hotel. More than 800 delegates to the 26th annual Intersociety Energy Conversion Engineering Conference filled meeting rooms in the hotel, but a mere scattering of attendees passed through the auditorium door labeled Innovative and Advanced Systems. One man muttered as he slipped out of range of Jeane's camera, "If my colleagues knew I was in here, I'd lose my position."

It became apparent why. Although the panel speakers onstage had doctorates in physics and mathematics, their discussions strayed outside the comfort level of some audience members.

Walter Rosenthal, a tall, lanky test engineer employed by a large aerospace company in California, responded when someone asked if anyone on the panel had seen a 'more output than input' machine working. Rosenthal replied, "I have." In his careful, unhurried manner, he said he had witnessed a device called the Vacuum Triode Amplifier.

"I have personally seen Floyd Sweet's machine operating. It was running that day at about 500 watts *output*...running small motors...It was jump-started with a nine-volt battery. There was no other electrical input required to the machine once it was started. I saw it started and stopped repeatedly in a two- to three-hour time period. There was no connection to the power line whatsoever."

Rosenthal answered questions from the audience. No, there were no moving parts, and no electrical connection from the Sweet machine to any other power system. He had seen Sweet's invention putting out useful electricity with only a tiny input from an electromagnetic signal going into a specially-conditioned block of barium ferrite (a hard material containing iron; it can be magnetized).

A Cal Tech delegate said the comments were outrageous. "To present such a remark at an engineering conference is the height of irresponsibility! It violates virtually every conceivable concept known to engineers!"

The test engineer invited the indignant academic to view a video of Sweet's invention, but the man stalked away, up the aisle and out the door.

When Jeane telephoned Floyd Sweet weeks later, the elderly inventor was willing to talk. He had worked at a General Electric plant and earned the nickname "Sparky" when he was young, after he misconnected wires and an instrument exploded in a spectacular display of sparks.

In his job, Sweet on rare occasions had seen an effect called self-oscillation occur in electric transformers. He felt it could be coaxed into doing something useful, so he looked for a way to disturb a magnet's field to cause the field to continue to shake by itself.[133] It would be similar to striking a bell and having the bell keep on ringing.

Sweet wondered if a small amount of universal energy could be captured to set a magnetic field in motion. A constantly moving magnetic field gives you electricity when it's near a coil of conductive wire such as copper. It induces current to flow in the wire.

What if the magnet stays still but you set its magnetic field in motion?

Sweet believed the entire universe is permeated with a magnetic field. He would have happily spent all his time building a device to tap into that when he retired and moved to California. His wife Rose became an invalid, so instead he worked on his device when not tending to her needs.

Experimenting, he came up with a set of specially conditioned magnets wound with wires. As a test he discharged electrical current into a coiled wire, then connected a twelve-volt lightbulb to the coil. The bulb would light if the device produced electricity.

---

133 Manning, Jeane, *The Coming Energy Revolution*, Avery Publishing Group, NY, 1996, pp. 71-79.

There was an unexpectedly bright flash. The bulb received so much power that it melted. "Sparky" Sweet recalled that Rose had called out from another room, "What did you blow up now?"

'What did you blow up now?'

The inventor had returned to his workbench to make further models. Needing a theory to explain the dazzling light, he remembered hearing physics theorist Lt. Col. Thomas Bearden and electronics expert John Bedini on a local radio show talking about unorthodox concepts. Sweet called Bedini, who arranged for Bearden to visit Sweet.

Bearden saw the curious device seemingly pulling power out of the air; only a tiny fraction of a watt went in to start it. After testing Sweet's assembly of magnets and coils, Bearden named it the Vacuum Triode Amplifier (VTA). Bearden decided that the device served as a gate through which universal energy was herded into an electric circuit.

Bedini supported Bearden's report. Bedini saw Sweet's device started by a very tiny signal that triggered an electrical output at least four times as powerful.[134]

Instead of creating heat as electric equipment usually does, Sweet's VTA did run cold. Its insides were as much as twenty degrees cooler than the surrounding air. The more electrical load on the device, the cooler it became. When VTA wires were accidentally shorted out, Sweet said, they flashed with a brilliant burst of light and became covered with frost.

Development of the device was slowed by trouble with materials and processes, and by financial entanglements. As with first models of any new technology, the VTAs Sweet built were unreliable. At times their output went down at night and picked up again during the day. Sometimes, they just stopped working for no apparent reason.

Early in the VTA's development, someone broke into the apartment and stole Sweet's laboratory notes. He then began to code his notes. After further harassment, Sweet temporarily stopped work on his invention out of concern

---

134 www.energyfromthevacuum.com/Disc30Sweet_Memories/Disc30SweetMemories. html.

for his ill wife. "They must have known I stopped; they didn't torment me anymore," he told Jeane.

Floyd Sweet suffered a fatal heart attack in 1995 at age 83. The VTA intellectual property was bogged down in legal problems over Sweet having signed conflicting agreements with financiers.

Sweet had frustrated fellow researchers by keeping secret his most important process—how he conditioned magnets that are the heart of the VTA. Did he pump the magnets with powerful electromagnetic pulses to shake up their internal structure? He refused to give details.

Despite the extreme secrecy, the fact that other qualified engineers witnessed his VTA putting out much more electricity than the tiny starting input indicated it is possible.

MANELAS' SUCCESS AMAZES EXPERT

Jeane followed the more recent development, Arthur Manelas' invention, through conversations with Brian Ahern, PhD. He is a mainstream scientist whose trajectory was changed by meeting Manelas and seeing the oscillating power source that recharged batteries for more than twenty months.[135]

Ahern has a doctorate in material science and is an expert on nano-sized materials. He and his colleagues discovered unique vibrational properties in materials in a size range of about three to 12 nanometers (a nanometer is a billionth of a meter.) That discovery eventually required him to investigate the field that had been mis-named cold fusion and renamed Low Energy Nuclear Reactions (Chapter 20). To his surprise, he had found gems of truth in LENR claims. It opened him somewhat to topics beyond the mainstream.

Dr. Ahern knew nothing about the late Floyd Sweet until the end of August 2011. Ahern was in his own laboratory in Massachusetts doing a study for the Electric Power Research Institute when the high-voltage power supply he was using burned out. Luckily, a specialist who could fix it arrived from the neighboring state of New Hampshire.

---

135 Ahern, Brian, *Conference on Future Energy*, Albuquerque, NM, August 11, 2018.

That specialist was Arthur Manelas. Mutual acquaintances had alerted him that Dr. Ahern too was discharging high-voltage electricity into nano-materials. Manelas took Ahern's damaged power supply back to his workshop in New Hampshire.

When Dr. Ahern drove to Manelas' home to pick up the repaired equipment, the men talked easily. Manelas' own varied career had included starting a solar company and patenting solar technology.

One of his inventions was revolutionary.

Manelas showed Dr. Ahern his 1997 Solectria electric sports car. Its batteries were charged by a system based on Tesla's and Sweet's ideas. Dr. Ahern couldn't see where its power would come from, yet Manelas said it worked continuously.

At hearing that, another mainstream scientist might have jumped back into his own car and laid tire tracks on the man's driveway. Dr. Ahern, however, was willing to look for an explanation of what Manelas claimed.

Dr. Ahern decided to design tests for the device and its battery pack, such as running the system outside of the car on a bench top. They would cage the equipment to isolate it from outside electromagnetic input.

With Manelas' permission, Dr. Ahern recruited other talented engineers to help investigate the system including the flat slab compressed from nano-sized ferrite particles.

Seen in Manelas' revolutionary battery-charger: The four-by-six-inch billet (slab) was made from strontium ferrite,[136] It had unusual magnetic fields, and was the core of a transformer in the electrical circuit of the device. Magnetic wire had been wound around the billet in three directions. High-voltage electricity was pulsed through two of the windings, which somehow created excess power that was extracted from the third winding and charged the car's battery pack.

They drove the car 25 miles with four passengers. The battery capacity increased from 68 per cent when they started to 89 per cent. Dr. Ahern

---

136   One account said it was an inch thick; another said a half-inch.

later said he is certain about those measurements but doesn't know why it increased.[137]

They let the car sit for a week, completely encapsulated and isolated, with data-logging voltmeters attached to the batteries. It recharged itself "which is preposterous," Dr. Ahern reported to a gathering of scientists. "It seemed like a violation of the first law of thermodynamics."[138]

Data-logging recorded substantial dips in energy at certain times. The engineers later found an exact correlation between those dips and the timing of large aurora borealis (Northern Lights) events. They were baffled by the interaction between the device and the magnetosphere that surrounds Earth.

COOLING INSTEAD OF HEATING

There was yet another paradigm-busting discovery and it echoed Floyd Sweet's. To measure temperature the scientists placed several thermistors around the trunk of the car when the system was running in it, and another thermistor on the billet, the specially magnetized bar of metal. During testing, the billet stayed five degrees Centigrade below the temperature of its surroundings, despite technical specifications saying the core of a transformer should heat up to about five degrees warmer than the surrounding air temperature when working.

"We don't understand it," Dr. Ahern said. "This is new physics."[139]

In light of what he knows about what happens on the nano-scale, Dr. Ahern speculated that tiny magnetic vortices suck in energy from the so-called vacuum of space as they collectively collapse then re-form. Publicly, he only says he doesn't know the source of excess energy. He told us that relatively few people know about "cooperative oscillations" among nano-grain ferrite particles in a certain range of particle size where magnetic vortices arise

---

137 Ahern, Brian, "Nanomagnetism for Energy Production," summarized in Infinite Energy magazine, May/June 2014, p. 18.
138 Ahern, Brian, Cold Fusion/LANR Colloquium, March 21, 2014, MIT, Cambridge, Massachusetts.
139 Ahern, Brian, Ibid.

like miniature whirlwinds. He said those particles operate by a different set of rules.

Arthur Manelas and his family wanted to work on the project themselves for the next year. Dr. Ahern later reported to an audience of scientists that Manelas had refused to take any investment money, even from a willing group of potential investors. But about nine months into the next year, Manelas sent an email telling Dr. Ahern that the time had come "for you to take over, Brian" and for investments to move the project forward.[140]

TRAGIC TIMING

Two days later and before Dr. Ahern could begin to learn more about the invention, Arthur Manelas suffered an aneurism. After the stroke, he never recovered his mental faculties. He died without leaving a circuit diagram or notes about how his device operated. And he had taken his device apart to rebuild two new ones.

However, other engineers had seen how well it had operated during testing in 2011-2012. Dr. Ahern hypothesized that the excess power comes from a new magnetic interaction.

Arthur Manelas did not get a chance to fulfill his intention to share his knowledge, but his story is not a repeat of the Floyd Sweet saga. Manelas left a legacy that is more about collaboration than competition or fear.

After Manelas' obituary reached the Internet, seasoned researcher Mark Dansie emphasized that highly qualified technical experts worked with Arthur Manelas and continue the work in their own laboratories. Many groups and individuals are involved, Dansie wrote, and each in their own time will reveal or publicize whatever they develop. Many share information within their networks.[141]

The results of keeping a low profile, as Arthur Manelas wisely had done, are distinctly different from results of being excessively secretive.

---

140   Cold Fusion/LANR Colloquium, March 21, 2014, MIT, Cambridge, Massachusetts.
141   Dansie, Mark, http://Revolution-Green.com.

CHAPTER 16
THE NEW BUZZWORD

Etheric science is expansive, cosmic, metaphysical,
spiritual—and way more fun than quantum...

—George Trinkaus, publisher, High Voltage Press.

IN THE CLASSIC FILM *THE GRADUATE* THERE'S AN IRONIC moment heightened by 1960's attitudes. The character played by actor Dustin Hoffman receives advice about "the next big thing" for a career. His parents are celebrating his graduation from university with a poolside party. One of their friends hustles the confused graduate aside to reveal a direction the young man can take. The man smiles benevolently as he tells the graduate.

"Plastics."

That word was underwhelming to the graduate and is a recognized problem today. The plastics industry has made fortunes and useful products, but plastic crumbles into bits that fish, wildlife and soil microorganisms mistake for food. Durable plastic garbage gathers into artificial islands in oceans, and birds die with stomachs full of plastic.

The parental generation today might well advise a graduate with a word that doesn't need to have anything to do with fossil fuel by-products.

"Plasma."

Plasma isn't something to be bought. It is one of the four common states of matter, an ionized gas. To picture the four states, consider a cup of steaming hot drink. The cup rigidly holds its shape; it's *solid*. The drink is *liquid*. The air you blow to cool the drink is a *gas*. *Plasma* is in the fluorescent tubes lighting the room.

When a burst of electrical power tears liquid or gas atoms apart to create a plasma, electrons and ions are freed to work together in ways that they can't in other states of matter. Plasma is the transformative part of some small but powerful convertors of electrical power.

BALL LIGHTNING: PLASMA IN ACTION

Nikola Tesla was surprised to see fireballs among streamers of sparks from his high-voltage experiments in his Colorado Springs laboratory. Fireballs were only a puzzling nuisance, but he followed his curiosity and eventually claimed to make ball lightning deliberately.[142]

Ball lightning is spooky. Over the centuries individuals have sighted glowing spheres that float slowly parallel to the ground or down the aisle of an airplane or pass through objects without heating anything. The ephemeral balls of cold fire were rare and therefore frightening.

The mystery is where they come from and how excited particles create a luminous ball that survives for eight or more seconds. A standard plasma reaction would be so hot it would heat the surrounding air and rise, not skim the ground.

Ball lightning is a free spirit. Instead of being confined in a fusion reactor, it appears out in the open and stabilizes itself for a surprising number of seconds before disappearing.

A German study group made ball lightning-like plasma clouds with luminous circular spirals at diameters of ten to twenty centimeters. Inside the spirals they saw a cloud of ionized gas that they named a "quantum nucleus." The group was reported to be stunned at the potential gain of energy, but a piece of paper placed above the plasma didn't burn.[143]

---

142   Tesla, Nikola *Colorado Springs Notes* p. 368-370.

143   http://newenergyandfuel//com/2008/09/16/power-from-ball-lightning.

CHUKANOV DEVICE'S EXCESSIVE POWER

Kiril Chukanov, PhD, a thermodynamics specialist originally from Bulgaria, created ball lightning in a quartz sphere within an industrial microwave oven and decided that its unusual electrical features could produce useable energy. Decades of experiments yielded dramatic results. When photos of his later work in a laboratory in China came to our attention, he had worked with five other specialists there for two years. To update us, Dr. Chukanov emailed from Bulgaria:[144]

"…For the first time in my long work on my discovery, almost thirty years, I created ball lightning in very dense media… the over-unity (excess energy) is very big—tens and hundreds of times."

He had nearly started a business "but my Chinese investors had another plan…Without me, however, they cannot solve the problem with industrial QFE (Quantum Free Energy) generators." At the time he said he was content to give the Chinese nation his technology because his best work on the generators was in Canada and China, having received funding in those countries.

The plasma processes of Dr. Chukanov and Dr. Randell Mills (Chapter 20) create a similar challenge, how to work with such extremely bright light. Instead of the common problem of not enough power, they over-achieve. The inventors each also developed unique theories.

UNIVERSE IS ELECTRIC

Plasma is active in our sun and the cores of stars and in quasars, pulsars and supernovas. The book *Thunderbolts of the Gods*[145] popularized an electric universe theory that describes plasma electrically interconnecting the universe.

On Earth, plasma acts in flames, lightning and the auroras. Mysteries of our weather can be solved if plasma's role is taken into account.

---

144   Chukanov, Kiril, emails to Jeane Manning, August 2017.
145   Talbott, David and Thornhill, Wallace, *Thunderbolts of the Gods.*

Plasma also features in your energy future. Since a plasma discharge acts as a whole instead of as a random bunch of atoms, experiments indicate it's a powerful and capable state of matter for interacting with universal energy.

Greek mathematics professor Panos Pappas, PhD, said that sparks from spark plugs can interact with universal energy. More recently in Washington state, Aaron Murakami developed a plasma ignition method that puts out a brilliant white ball of light in a spark gap.[146]

Is tapping ether energy from the cosmos via small plasma-based machines the next big leap for humankind?[147] Conventional scientists instead try for electric power by making a plasma reaction so hot it fuses atoms together. Dissident scientists question whether that is really the way the universe works in our sun and the stars.

Unlike the well-funded "big science" particle accelerators, the American company Lawrenceville Plasma Physics (LLP)[148] and several other companies have methods that could lead to fusion without radioactivity. LLP chief scientist Eric Lerner cites discoveries in astrophysics that link what he sees on a small scale and vast processes throughout the cosmos.[149] Electric currents twist into rope-like filaments confined by magnetic fields.

Curtains of plasma filaments are seen in the northern lights' sheets of current. Tails of comets and our sun's coronal streamers are plasma filaments. We do live in an electric universe, not a mere vacuum of space with bits of matter.

SMALL PLASMA DEVICES FOR HOME ELECTRICAL NEEDS?

For small-scale ways to work with plasma, experimenters with modest funding are eyeing the *Papp engine*, reported to have produced at least ten times more

---

146 Murakami, Aaron, http://emediapress.com/aaronmurakami/hackingtheaether.
147 http://ignitionsecrets.com.
148 Review Committee Evaluation of the Lawrenceville Plasma Physics Focus Fusion program, Nov.28, 2013.
149 LPP Fusion Report, Apr. 24, 2018.

power output than input. (Papp is pronounced "Pop" in Josef Papp's country of origin Hungary.)

That cool-running engine of more than 100 horsepower—75 kilowatts—ran on a mixture of noble gases. Noble gases usually don't react, but in the Papp engine they exploded violently and powered a device retrofitted from an ordinary gasoline engine. It ran without needing a cooling system, fuel system or exhaust. There was no dangerous radiation or waste.

JOSEF PAPP MYSTERY

The late Josef Papp was an ex-pilot who left Hungary after a Soviet invasion. He built an engine that amazed engineers who saw it running in closed rooms for hours. The engine had huge turning power even at low revolutions-per-minute.

The late Eugene Mallove, PhD, wrote in *Infinite Energy* magazine that Papp's engine appears impossible until the evidence is examined. "Once the battery-driven electric starter revved up the Papp engine, according to dozens of initially-skeptical witnesses, the engine—equipped with an alternator—ran with no outside electrical input. And even if such 'miracle' reactions of noble gases should produce interminable explosions from a tiny volume of gas pushing pistons and driving a large flywheel, why didn't such an engine run very hot? It didn't."[150]

Dr. Mallove said that only a radical departure from conventional under-standings will explain that engine. He expected that the future scientific paradigm will explain interactions that are manifestations of an ether physics.

---

150   Mallove, Eugene, *Infinite Energy* Sept./October 2003.

CHAPTER 17
SEEKING 'ZERO-
POINT' ENERGY

As the paradigm shifts, the special interests that went into
suppressing the discovery will then produce enormous
investment capital to develop it...by uplifting the world,
we all win.

—Moray King, system engineer. [151]

THE ENERGY REVOLUTION IS GATHERING SPEED. NO ONE
will need to spend 44 years uncovering keys to a new energy source as Moray
King has done. He synthesized the work of other theorists and of inventors
and found a master key—in water.

When his epic search began, King was a graduate student in the Masters/
PhD program for system engineering at the University of Pennsylvania.
He held the standard opinion that the "vacuum" of space was empty, since
Einstein's theory of relativity didn't need an ether. [152]

His roommate, a graduate student in physics, one day handed King a book
titled *Gravitation,* [153] describing how space responds to tiny fluctuations of its
fabric. Quantum physicists had named those fluctuations zero-point energy.

---

151   King, Moray B., *Tapping the Zero-Point Energy: How 'free energy' and 'antigravity' might*
be possible with today's physics, Paraclete Publishing, Provo, Utah, 1989, page 168.
152   Ibid. , page i.
153   Misner, Thorne, and Wheeler, *Gravitation*, W.H. Freeman, NY, 1970.

Electrical engineering students he knew had never heard of zero-point energy. He asked engineering professors, they hadn't either.

King eagerly read any physics literature in his university's library that mentioned zero-point energy. He was amazed to learn what happens on a tiny scale twenty orders of magnitude smaller than an electron.

Zero-point energy from a higher dimensional space was said to enter and exit our three-dimensional space. Physicist John A. Wheeler, PhD, said those movements create what he called quantum foam. Dr. Wheeler's model of space was a vastly more dynamic and turbulent ether than the 19th century concept.

As engineers do, King asked a practical question. "How can we harness this zero-point energy?"

He persisted. Even after finishing his course work for a PhD he took more courses in quantum mechanics and questioned professors. Why weren't they looking at zero-point energy as a possible energy source?

ENTROPY RULES IT OUT?

The professors assumed that zero-point fluctuations just even out. They quoted the second law of thermodynamics about *entropy*, the belief that everything is doomed to increasing disorder until it comes to a dead rest. Traditional science said that universal or zero-point energy can't do practical work because it is random motion.

Russian scientist Viscount Ilya Romanovich Prigogine, however, proved that randomness could organize into order. He showed that under certain circumstances a chaotic system can suddenly organize itself into a different status. Dr. Prigogine won the 1977 Nobel prize in chemistry for that discovery. It expands the second law of thermodynamics.

King took note of the conditions Dr. Prigogine said are needed before a system will self-organize. It must be non-linear, far from equilibrium, and have an energy flux (flow) going through it.

In a nonlinear system the change in an input can be totally different from the change in what comes out. In other words, the change to the output is not in proportion to the change of the input; it's unpredictable. A plasma is a nonlinear system.

Prigogine's conditions for self-organization could apply to the quantum foam, King realized, if an abrupt discharge of electricity jerked ions (atoms that are electrically charged from having lost or gained electrons) and pushed a plasma into a far-from-equilibrium state.

(A 1968 theoretical paper by Elizabeth A. Rauscher, PhD, as far as we know was the first to suggest that zero-point energy can cohere in plasma.[154])

King had to propose a topic for his PhD thesis, so he boldly wrote *Tapping the Zero-Point Energy* and suggested that ball lightning might create order out of random fluctuations. Being an outgoing personality and having made friends with professors helped him at this point. Some professors were open to considering his topic—if proven by an experiment.

His life changed when scientists at a research institute in Utah saw his writings and were intrigued at the prospect of doing experiments. King accepted their job offer to work in industry.[155]

In his spare time, he spread his message among inventors: "Only a repeating experiment will cause a paradigm shift. It must be simple because there's no funding for an experiment that violates the existing paradigm." The first book he wrote ended with that call to action, adding "there is tremendous joy and fulfillment for those who freely share to uplift an entire planet."[156]

King's enthusiasm was renewed whenever a physicist published a peer-reviewed paper about how zero-point energy creates forces such as gravity and inertia.[157]

---

154   Rauscher, E.A., (1968), "Electron Interactions and Quantum Plasma Physics," *Journal of Plasma Physics*, 2(4), p. 517.

155   Eyring Research Institute hired Moray King.

156   King, Moray B., *Tapping the Zero-Point Energy: How free energy and 'antigravity' might be possible with today's physics*, Paraclete Publishing, Provo, Utah, 1989, page 169.

157   Puthoff, H. E, "Gravity as a zero-point Fluctuation Force" *Physical Review A* 39(5),2333, 1989.

T. H. MORAY'S 50 KILOWATTS OF FREE ENERGY

When Moray King encountered the book *The Sea of Energy in which the Earth Floats,* he noticed the author's surname Moray was identical to his own unusual first name.[158] Further, T. Henry Moray (1892-1974) had also lived in the state of Utah and had invented a paradigm-changing energy device. It got his attention.

Technical experts had witnessed T. Henry Moray's device putting out fifty kilowatts of electricity without any known input power. It couldn't be explained by standard scientific theories. The electricity from it had a strange characteristic called cold current;[159] thin wires guided so much power that ordinary electricity would have melted the wires.

After King researched the inventor's life, he saw it as a tragedy. "...technical success was followed by business subversion, government corruption, threats and assassination attempts." [160]

Despite threats, T.H. Moray repeatedly demonstrated his strange electric generator to credible witnesses. The only threat that stopped that was advice from his patent attorneys. Under patent laws he could lose his rights to a patent if he continued to show his invention to everyone.

The U.S. Patent Office rejected seven patent applications for the Radiant Energy Device on grounds that it did not fit the known physics. And his semiconductor technology was so advanced that the patent examiner rejected its patent application because the examiner could not see how the device could work.

In the most destructive incident, a man took a sledgehammer or, some reports say, an axe and smashed Moray's Radiant Energy Device. Whatever the motivation, the attack wiped out years of development work and damaged beyond repair some of the device's most vital parts. Instead of rebuilding the

---

158 *Moray, Thomas Henry, The Sea of Energy in which the Earth Floats, first self-published in 1960, Utah.*

159 At a meeting of 'HHO' researchers in Maryland, *Jeane Manning witnessed a demonstration of cold electricity from a system using Nikola Tesla's "hairpin circuit."*

160 King, Moray, B. *The Energy Machine of T. Henry Moray: Zero-Point Energy and Pulsed Plasma Physics Adventures Unlimited Press, Illinois, 2005, p.29.*

machine, King reports, the inventor focused his research on another anomaly that happened in his device's plasma tubes—transmutation of elements.

KING LOOKS OUTSIDE THE BOX

King found that academics generally don't take zero-point energy beyond theorizing. Looking elsewhere for a world-changing experiment, he began presenting his message at non-conventional energy conferences such as the International Tesla Society. He freely gave technical tips to "garage experimenters" and hobbyists.

By the time that he wrote a book about Moray's sophisticated free energy device, King had additional insights.[161] For instance, Moray had discovered that abrupt pulsing to maintain oscillation of ions in plasma is a key activator for getting anomalously large amounts of energy. (To oscillate means to swing back and forth or to change from one extreme to another with a steady rhythm.)

FROM THE VORTEX TO THE TORUS
AND BALL LIGHTNING

King was fascinated by how a vortex formed in plasma could spin into the shape of a fat donut called a torus. (Film maker Foster Gamble independently envisioned the torus phenomenon and began his own journey of discovery about non-standard energy inventions, resulting in a popular documentary.)[162]

Over the years King also kept his eye on devices called electrolyzers that hobbyists built to break up water into hydrogen and oxygen. Their goal was to inject hydrogen into the fuel-burning in their car engines to get better mileage. Some said they could run an engine or electric generator on only water. They were encouraged by the story of Stanley Allen Meyer of Ohio

---

161   King, Moray B., Ibid.
162   Foster and Kimberly Gamble co-produced the documentary *Thrive*.

(1940–1998) having run a dune buggy on tap water. However, Meyer did not reveal all his secrets in his patents. Meyer once told Jeane that ether or zero-point energy did enter into his water fuel system. He did not say at what point.[163]

Stanley Meyer, American inventor (Jeane Manning photo)

Critics are correct in that it is impossible to run a car engine on hydrogen gained from zapping water with electricity using standard electrolysis. How could the occasional experimenter be getting anomalous extra mileage or even running an engine on mainly water? A few years later King learned how.

Separate discoveries of electronics pioneer Ken Shoulders and Russian scientists led to King's breakthrough in finding a key to a new energy source.[164]

---

163 Note for technical replicators: An experimenter recently told us that Stan Meyer apparently insulated his electrodes so the electricity did not make direct contact with water. Therefore Meyer's invention was a "field-effect device, not current-operated. It was not electrolysis," our source said.

164 King, Moray, *Water: The Key to New Energy,* Adventures Unlimited Press, Illinois, 2017.

CHAPTER 18
THUNDERCLAP POWER

The technology coming out of vacuum energy, zero-point energy, spacetime energy—call it what you will—is going to take a while. In the meantime we've got to start changing the way we live.

Author and aerospace journalist Nick Cook[165]

KEN SHOULDERS WAS AN INDIANA JONES OF NEW ENERGY. Like the fictional Jones, Shoulders was a resourceful adventurer possessing both brainpower and attitude. He was irreverently independent. However, instead of hacking through tropical jungles, Shoulders blazed trails by exploring beyond the frontiers of known science.

Using sophisticated laboratory equipment acquired during his industrial career, he did what mainstream scientists said is impossible—making and controlling stable tightly-packed clusters of 100 million electrons. In short, he learned how to create something tiny but powerful. His microscopic clusters of condensed electric charge put out more than thirty times the energy required to produce them.

Shoulders said the electricity involved is static electricity as in the spark that snaps from a doorknob if you drag your feet across a carpet.[166] A charge cluster

---

165   Nick Cook, author of *Hunt for Zero Point*, interviewed by Jeane Manning, Vancouver BC, 2007.
166   Ken Shoulders interviewed by Jeane Manning.

is like a miniature form of ball lightning. The charge cluster gains stability from the way it creates itself from a helical filament of electrical charge.

Its vortex shape evolves quickly into a circulating vortex ring of plasma ions and electrons known as a *plasmoid*. At that point it could look like a microscopic "slinky" toy closed in on itself in a donut shape. Its self-organizing toroidal motion makes the miniature ball lightning stable for seconds of time, long enough to impact something.[167]

Ken Shoulders, pioneer in electronics (Jeane Manning photo)

Textbooks say electrons shouldn't cluster; similar electrical charges repel each other. Shoulders tried to find out what supplied the large amounts of energy needed to bring electrons together. He eventually named the clusters EVOs, Exotic Vacuum Objects after he speculated they get their extra power by cohering the universal energy.

It wasn't his first time as a trailblazer. Ken Shoulders developed much of today's microcircuit technology. He was at Stanford Research Institute International for ten years as a staff scientist and pioneered a field called vacuum nano-electronics. While there he built specialized instruments, inventing tools he needed. He also had a non-teaching staff position at Massachusetts Institute of Technology (MIT).

Shoulders' research isn't owned by any institution; he explored the high-density charge cluster in his self-funded laboratory.[168]

He called his charge clusters "little engines of vast complexity…" What inspired his awe about the tiny entities was that as they self-organized they

---

167   King, Moray, *Ibid.*

168   Ken Shoulders discovered EVOs by accident. Around 1980, physicists at the Stevens Institute in Hoboken, New Jersey, introduced him to strange strings of particles—vortex filaments. He found they were about as broad as they were long but looked like strings to most researchers because the researchers couldn't stop the motion of the extremely fast-moving blobs. Shoulders learned how to get clear pictures and saw that the "strings" are little beadlike structures. Through relentless experimentation, he created conditions under which electrons join together into remarkably stable ring-shaped groups, often looking like a necklace of tiny donuts.

almost seemed to have an intelligence. They formed into various sizes but were uniform in organization and behavior. "It's some law of nature that's just not spelled out for us yet."[169]

'THE COMPUTER ATE IT'

No other inventor had succeeded in getting a patent whose claim was based on zero-point energy, but Shoulders broke through with his 1991 patent titled Energy Conversion Using High Charge Density. Shoulders was forced to be strategic, said a science writer in Utah. The late Hal Fox, PhD said Shoulders outwitted officials who want to slap a secrecy order on revolutionary energy breakthroughs.

"The biggest stopper occurs at the U.S. Patent Office, where each incoming patent application goes to a government agent responsible to classify things that supposedly are of strong national interest," Dr. Fox wrote.[170] Protecting the interests of the oil industry is apparently considered synonymous with national security.

"The government should have no right to classify work that was done without the use of federal funds," Dr. Fox wrote. "Therefore, if one wants to ensure that his/her patent is not classified, one has to mail out or distribute widely the information as soon as the patent application is mailed."

"Shoulders' first patent application was immediately classified 'secret', but his patent attorney had worked with (wealthy Texan entrepreneur) Bill Church who paid the money, and Ken. They wrote a book and mailed it to scientists in over 25 countries the day they filed the patent."

"The patent office asked for the mailing list, but 'the computer had eaten it'… Three days and $10,000 later, the patent office gave up and removed the classification."

---

169   Ken Shoulders in 1994 interview with Jeane Manning.
170   Hal Fox private correspondence March 29, 2001.

GLIMPSING A FUTURE

Shoulders' well-equipped science laboratory was for years a few steps away from the home he shared with his wife Clare in a rural area north of San Francisco. There he continued to make breakthroughs, often working with their son Steve. What the two researchers saw under the microscope was another world and hinted of possibilities for future machines.

Shoulders conjectured that a charge cluster is a type of universal clutch as it grabs the basic fabric of the universe. He discovered that an EVO could self-accelerate to about a tenth of the speed of light.

His patent included unusual statements. "The EV may be considered as being continually formed as it propagates…the ultimate source of the energy appears to be zero-point radiation of the vacuum continuum."

Since it suggested the universal energy could be tapped, Dr. Fox said, the patent should have been worldwide news about a dramatic new scientific study.[171] One popular science magazine of that time did recognize Shoulders' advance. *Omni* featured an article and full-page photo of him.[172]

In the end Shoulders had to admit that he couldn't tame the electron entities. He tried to harness their excessive power by accelerating collections of charge clusters toward various target materials. The clusters destroyed every target and created shock waves that smashed into surrounding electronics. Instead of too little power, they had too much.

Until felled by cancer in the summer of 2013, Shoulders was an explorer. He let others worry about whether his discoveries fit within accepted boundaries of scientific theory.

MORAY KING PICKS UP THE TRAIL

Moray King was excited about Shoulders' discoveries and became friends with the electronics genius. Russian scientists who independently discovered clusters of charge also caught King's attention. That validated Shoulders' strange

---

171   Hal Fox, Climates of Change Conference, University of Victoria, BC, 2000.
172   Davies, Owen. "Volatile Vacuums," *Omni* February 1991, 50-56.

electron effects, but it was not the experiment thousands of experimenters could do in their garages that King hoped to find.

He saw similarities between what Shoulders' EVOs could do and strange effects documented in the gas named after Bulgarian-Australian experimenter Yull Brown. King credited the Canadian inventor George Wiseman for being the first to propose that in addition to hydrogen and oxygen, "Brown's gas" contained a water form that Wiseman named "electrically expanded water."

In experiments with either that gas or EVOs, hard metals vaporize without much heat involved. And you can pass your hand through the fairly cool flame of Wiseman's welding torch fueled by Brown's gas. The flame doesn't boil water, yet it cuts through tungsten. That metal usually has a melting point of 6192 degrees Fahrenheit (3422 Celsius).

King says as the unusual gas apparently implodes it traps a coherent form of energy. It fortunately reacts with metal materials differently than with flesh.

Transmutation of elements is another anomalous effect documented in Shoulders' EVOs, Brown's gas experiments and the imploding bubbles made by a company called NanoSpire.[173] North American mainstream scientists shun such claims because they believe only dangerous nuclear reactions can transmute metals, but Russian scientists study it openly. [174]

Other scientists also see elements change.[175] A Japanese physicist wrote a book *Nuclear Transmutation: The Reality of Cold Fusion*.[176] Sochi, Russia, hosts meetings on ball lightning and cold transmutation. Experiments have disabled radioactivity, yet it is wisest to stop making those wastes.

---

173   Eagle-Research.com.

174   The Proton-21 laboratory in Kiev, Ukraine, hires university professors to study trans-mutation of elements such as copper due to plasmoid strikes.

175   Vladimir Vysotskii, PhD is a leading scientist in transmutation field. ColdFusionNow podcasts such information in the English language.

176   Mizuno, Tadahiko, *Nuclear Transmutation: the Reality of Cold Fusion*, Cold Fusion Technology, 1998.

PURSUING THE CLUES

Shoulders needed perfectly round blobs of liquid metal to create his charge clusters, because a sphere seemed to be a template for the plasma discharge to for m around. The metal spheres the Shoulders created by a high-voltage discharge were tiny, mere millionths of a meter. A sudden electrical discharge causes each sphere to dimple into a torus like a microscopic donut.

King had also learned that the extremely high speed vortexian spin of those toroidal plasmoids could pull in zero-point energy. He thought about hobbyists who had gained power in engines from the gas they called Brown's gas. King knew that the source of extra power had to be more than hydrogen, because electrolysis of hydrogen is not efficient enough to make that much difference in an internal combustion engine.

King gradually put together what had been happening in other experiments involving water and high voltage.[177] Instead of a microscopic molten metal blob, a very small water bubble or droplet could be the spherical template to guide the formation of an EVO if the droplet was stable enough to stay together when hit by the electrical discharge. The hobbyists who succeed in running an engine on what they think is hydrogen could instead be creating EVOs without knowing it.

THE WATER MIST CLUE

When he studied the dynamics of thunderclouds, King realized that a major clue to how to get extra energy is found in a mist of water particles so small they are nano-scaled (measured in billionths of a meter.)

Aaron Murakami in Washington state was independently demonstrating an anomaly that gave another clue. Murakami had been spraying a mist of water into his own very efficient plasma spark plug. In the mist, what had been just a respectable spark became a large, extremely bright flash of light. Each

---

177 MIT scientist Dr. Peter Graneau's water fog experiments were replicated by Gary Johnson, Ph.D.

spark event made such a loud bang that Murakami needed ear protection. It sounded like a sharp thunderclap.

By the time that Moray King gave a talk at the 2016 Energy Science and Technology Conference in Idaho, he had put clues together. The hobbyists' electrolyzers were not getting their excess power from hydrogen, so he figured that in a combustion chamber, fog particles are converted into microscopic ball lightning.

He told the audience how to insure that will happen: use a spark plug like the one Murakami invented. It releases a wide swath of plasma at the moment of an abrupt electrical discharge. Then briefly thousands of tiny "plasmoids" accelerate themselves because they are cohering zero-point energy. For a few milliseconds after each time the spark plug fires, the force pushing on the piston is huge, King said.

'ALL YOU HAVE TO DO IS MIMIC THE THUNDERCLAP EVENT, IN AN ENGINE.'

Without knowing about Moray King's ideas, in Florida inventor Walt Jenkins had already filed a patent on such a process.

King celebrates Jenkins' discovery and development of a 'thunderclap engine'; it is the example King has been seeking. King's recent book *Water: The Key to New Energy* gives the milestones of his four-decade long search for that understanding.[178]

---

178   King, Moray, *Ibid.*

CHAPTER 19
THE ARTIST AND THE
WATER ENGINE

Water, fresh or sea water, can now be processed into a primary fuel that replaces gasoline or diesel, for pennies per gallon.

—H2 Global LLC press release.

WALT JENKINS WAS AT HIS HOME IN FLORIDA YEARS AGO, ready for a diversion. He had finished producing a movie and needed to relax. Even on a fun project such as *Vampire Biker Babes*, a film producer can get burned out. (That was "a horror film but not X-rated or anything," he assured Jeane as he told the story of his invention.)

One of his diversions ever since he was a teenager has been tinkering with car engines, so he revisited earlier experiments. The goal was to obtain more hydrogen from electrolysis—electrically separating water into hydrogen and oxygen—for better gas mileage. Many hobbyists were trying to increase their car and motorbike mileage by mixing in a bit of hydrogen to make gasoline burn better. Jenkins succeeded in improving electrolysis beyond what textbooks said could be done.

That breakthrough was noteworthy, but insignificant compared to what he is doing today—running an engine on a combustible fuel created from 100 per cent water via a plasma process *without* electrolysis. The process creates a "thunderstorm" in combustion chambers by using highly-charged particles in high voltage fields.

Walter Jenkins,
inventor, artist

Jenkins sees himself as more of an artist than a technician, being a sculptor and painter as well a creator of motion pictures. However, without bothering to get a PhD or other degree from his years in university classrooms, he succeeded in powering an internal combustion engine on water. Long before that accomplishment, his technical talents had surfaced. During the decade that he worked in the Hollywood film industry, he came up with the concept of virtual sets. As a young man he was naïve about patents and disclosure of an invention, and others took the idea. After he left Hollywood, he engineered and built television stations as well as managed one in Orlando, Florida.

WHEN RIDICULE TURNS RIDICULOUS

During the early phase of Jenkins' water-as-fuel work, a typical incident occurred. We include his retelling of the dialogue because we often hear similar exchanges. And his story ends on a positive note.[179]

Walt Jenkins was visiting a solar power company in his home state of Florida when the company's managers were pitching to eight potential investors. He was not an investor, but had been invited by one of them, a Congressional candidate for whom he was producing video commercials.

An electrical engineer from the prestigious Massachusetts Institute of Technology (MIT) was impressing the group. He asked Jenkins what he does. This was before he had progressed beyond electrolysis.

"I told him. 'I've invented this hydrogen system. It extracts hydrogen out of water.' He (the engineer) told me, in a very doctrinal way, that I couldn't do

---

179   Walt Jenkins interviewed by Jeane Manning, June 2017.

what I was doing because I was violating the second law of thermodynamics. He knew nothing about what I was doing, yet he made that presumption."

Jenkins had challenged it. "That's not true at all. I'm using five amps (input to the electrolysis) on this car which I have in the parking lot, and then getting 72 miles per gallon. It's a Ford Windstar van, six-cylinder; factory standard is 19 miles a gallon."

The engineer told Jenkins to show him the vehicle. Jenkins demonstrated the device. The engineer again cited the second law of thermodynamics. Jenkins replied that the law does apply in many areas but not in all areas.

As he recounted the incident, Jenkins laughed easily. He recalled that when the engineer began goading him, the group at the solar facility "were kind of snickering, at me and with him, because he was MIT. Basically, treating me like I was a phony or something."

The engineer had said, condescendingly, "You can't get more energy out than you put it."

"That's not true either. You don't understand energy," Jenkins had replied. "Energy is like Tesla said—all around us, everywhere and abundant. The way you get energy to work for you is to create an imbalance."

When the engineer taunted, "I'll bet you believe in 'perpetual motion' too." Jenkins said he does, and cited examples of creating imbalance. One way would be by releasing energy stored in the earth by throwing a rock off a cliff and starting a 500-ton avalanche. Or releasing the thermal potential of a forest fire by flipping a match out a car window.

"The energy is always there; it's the *process* that releases it," Jenkins said. "If you don't believe that I'm legitimate and don't believe perpetual motion exists, I can prove it to you."

At that, the engineer had laughed, while others in the group echoed the derision.

Jenkins asked, "Tell me something. What are you made of?"

"Organic material," was the reply.

"I'll accept that. Now, what are you standing on? What is that made of?"

"Cement."

"The wall, what's that made of?"

"Concrete blocks." The engineer continued to express amused superiority.

"Tell me, what do they all have in common?"

The engineer seemed to be at a loss for words, Jenkins reported.

"I'll tell you what they have in common," Jenkins continued. "They all have atoms and molecules. And those, what do they have in common?" Jenkins' debate opponent was silent again.

Jenkins said, "They all move. They all have motion. And the galaxy, the universe, the entire everything we view, moves. Your math doesn't reach far enough to explain the dynamic power it would take to do that. Where does that moving force coming from? You can't even measure it. And you can't tell me anything that would explain that energy!"

The engineer admitted he did not know the origin of the force.

Jenkins said, "Well I know where it comes from."

He pointed to the sky, signifying God or the creator, and said, "It comes from Him. Or It. Whatever. Maybe it's a Her. I don't know. Maybe it's neither gender, maybe its just something out there. But it created the entire universe and it operates the entire universe and unless we understand its principles, we're playing around in the kindergarten class!"

The people at the solar facility stopped laughing at Walt Jenkins.

The aftermath to the incident was a pleasant surprise, revealing the MIT engineer's finer qualities. He approached Jenkins and asked, "Do you want to have coffee sometime?"

Jenkins appreciates the man's act of grace, the respect now given to an independent innovator. Some of Jenkin' encounters with highly paid consultants, however, have ended less graciously. Many experts seem to think that if an inventor doesn't have academic credentials in what he is doing, but claims a scientific breakthrough, he can't possibly have accomplished it and must be a fake and not worth visiting.

Such arrogance does more damage than merely angering inventors; it destroys their opportunities for being funded by the investors who pay those consultants.

Jenkins' experiences are similar to what other inventors such as the late John Bedini have gone through. For instance, a wealthy potential investor paid an engineering consultant to check out Jenkins and his water-fueled engines. The consultant talked at Jenkins on the telephone, insulted him, then refused to come to Florida because he was certain it would be a waste of time. Jenkins later found out that the consultant afterward sent a negative report about Jenkins to "the money guy." Without having seen the engines.

In addition to the common "You're wrong; the laws of physics say you can't do that," and "I don't know where you hid the gas tank; this just has to be a scam," and "You're not an engineer; how could you invent that?" Jenkins encountered other unhelpful attitudes.

He and a member of his team met with engineering department officials in a university whose endowment fund could have benefited from collaboration with his company. However, the response to the inventor could be summed up as "You don't have the academic training that I do; you need to be mentored by me." Experiences of other inventors indicate an unspoken part of the message: "so my university can patent your invention and I can publish a paper about it and then we're done with you."

Meanwhile, Jenkins does know highly-credentialed academics who arrive at a meeting with an open mind and willingness to help. Robert M. Haralick, PhD is a bright star in Jenkins' world, he adds. Dr. Haralick is a Distinguished Professor in computer science at the graduate center of City University of New York. Jenkins met the knowledgeable and delightfully irreverent Professor Haralick at the 2017 Energy Science and Technology Conference in northern Idaho.

SHRUGGING OFF PHYSICAL VIOLENCE

Jeane had seen online a mention of men having beat up the Florida-based inventor in an attempt to steal his technology. When she asked about it, Jenkins made light of the traumatic incident.

"Some guys in Tampa who were low-level." The thugs had claimed to be connected to the Mob, the mafia, but Jenkins did not believe that.

When the assault happened, he was still working in the electrolysis phase. His invention was strapped onto the front of his van—handy for adjusting the system. The Tampa group had promised to invest money in the research but kept stalling and instead pestered him about how it worked. Jenkins refused to give more information and said he wasn't interested in further association with them.

"It was all about greed."

The group nevertheless lured him into meeting in a parking lot in Tampa. When his back was turned, one of them beat Jenkins with a steel cane. The object was very heavy, with a big metal handle that looked like brass.

"This guy was beating me on the head with it until I was unconscious. Anyway, I went to hospital and that guy went to jail."

PUZZLING DREAM INSPIRES NEW APPROACH

Jenkins would rather talk about how he leaped beyond what other hobbyists were attempting. "Always be open to the process of thinking 'what if there's another way?'"

His early successes had caught him in the mental trap of thinking as if electrolysis was the only way. However, that process costs too much in energy input. And his goal was *not* to extract the energy out of water with electrolysis and make hydrogen and oxygen, he realized. He really wanted to make water into an energy source that works inside an internal combustion engine.

When he clarified his goal, he was free to abandon electrolysis and develop a radically different approach. A recurring dream reinforced his belief that there could be an unexpected solution.

At first, he didn't understand why he kept dreaming about a science demonstration that his eighth-grade teacher had given. The class was asked whether they could get more heat out of a brick of steel or from burning an equal sized block of wood. Students knew that you can't burn steel, so wood was the answer. The teacher then filed the wood and blew the sawdust across a blazing Bunsen burner to increase the flame. He did the same to the steel. Surprisingly, tiny bits of steel sparked an even bigger flame than sawdust. The haunting dream reminded Walt that new knowledge can be counterintuitive at first.

He thought about tiny particles of steel being made to burn, and "why not particles of water?"

Without knowing about Moray King's hypotheses, Walt Jenkins thought about where miniscule bits of water are found—in clouds. That line of reasoning led him to experiment with a dense fog mist, a spark plug, and an inch-long bolt of high-voltage, low-amperage electricity from a stun gun. It resulted in a brighter glow around the spark because the water droplets became a fuel.

TINY THUNDERCLAPS = BIG POWER

A lightning strike in nature can result in a thunderclap, but it doesn't explode the whole thundercloud, he realized. Natural conditions prevent that. Jenkins found that a different set of rules exists when tiny water droplets are injected or sucked into an engine's combustion chamber and are compressed by its piston while a plasma arc is created in the chamber. Then only a very small amount of carbon fuel is needed to power the engine.

His breakthroughs eventually reduced the amount of gasoline needed to be added to water to make a fuel, diminishing to five- or even two and a half per cent gasoline, mixed with water.

Jenkins' dune buggy is water-, solar- and wind-powered.
(photo courtesy of H2 Global)

Eventually he took an even more radical approach. "It doesn't just increase the mileage; it *replaces* fossil fuel," he told us. "Right now I can run an engine entirely on water fuel."

He has a proprietary way of conditioning water to make nano-sized droplets. Another key part of his process is the wide high-voltage spark plug he patented, inspired by Nikola Tesla's insight that a sphere holds electric charge longer than other shapes.

THE FOUNDATION

Jenkins' voice on the phone during an interview warmed as he outlined ideas for an angel fund, a foundation that would help people.[180] Any offer to buy the 'fuel from water' intellectual property from him would have to ensure a percentage of profits going to the foundation.

His habit of wholistic thinking had led to ideas for the foundation. He had been figuring out how to re-use old shipping containers, and also how to convert ocean-going ships to be powered by water as fuel. Jenkins chose a flexible solution—put his system into a recycled container. A large super-tanker would need four or five on deck, connected by hoses, to power the ship. A small tanker would only need one.

What else could he do with shipping containers? He realized that one containing a generator that runs on water could provide power for a small city or a village. It could be shipped wherever needed.

"I could take another one and make a medical triage unit, with robotics satellite-linked to medical teachings around the world. All you would need is an assistant in there, like an LPN (licensed practical nurse.) The robotics would do the rest. And you'd have a first-class surgery unit anywhere in the world."

"Then I thought about education. Make a satellite-linked kiosk, with computers linked to universities. So for some kid off in a far land, or disaster zone like Haiti when it had the earthquake, you could move these containers in there and put together energy, medical and education. You could bring

---

180   Ibid.

that education module, by linking it to satellite, all the way from K12 up to PhD. I thought, 'wow, that would really affect the world in a positive way... . I'm going to do it!'"

Jenkins wants ten per cent of the gross sale of his technology worldwide to go to the foundation. His determination to do that holds firm.

SHARK WATCHING

The stereotype of a vulnerable lone inventor does not describe Jenkins. The team at his company H2Global backs his humanitarian thinking. His network includes electrical, chemical and mechanical engineers, and PhDs in molecular and electromagnetic physics. In an open letter online, their senior consultant John Parker, PhD, said they had formed the network to help solve energy problems that threaten the environment.

Not everyone shares that dedication.

In one instance, a man who had become a billionaire in an automobile-related field was reportedly the money guy who sent a group to meet with Jenkins. They flew to Florida in a Gulf Stream jet airplane. Jenkins showed them the dune buggy and the generator running on water. The group must have been impressed, because they began negotiations. Jenkins insisted on keeping the majority ownership of his company, and said, "I will not allow this company to be bought just for the technology and buried, so don't even think about it."

Two weeks later, they returned with engineers who further checked out the technology. During hours of negotiation, the group offered up to ten million dollars, Jenkins recalls, and they wanted to own 70 per cent of his company.

"I said, 'If you think that's such a good deal, you take the 30; I'll take the 70. And you will give me a specific amount. If it's absurdly low like that, you can forget it.'"

They did not make a better offer, so Jenkins said goodbye.

That night, while the negotiator flew home, he phoned Jenkins. "He had imagined that I was going to go along with this deal," Jenkins recalls. "He said 'well, have you made up your head?'"

Jenkins repeated his earlier decision. "I respectfully decline your offer."

The negotiator "went ballistic" and yelled over the phone. "You just fucked up 500 million dollars!"

Jenkins was surprised. "Where did you get that figure? You never mentioned it."

He remembers the negotiator's fury. "I could have sold this company tomorrow for five hundred million dollars!"

"I said, 'So. You're the kind of guy who'd lowball me and turn around and sell it for 500 million. And you'd probably sell it to someone who would shut it down and bury it. That's exactly why I won't work with you. Goodbye!'"

In a separate incident, Jenkins came close to accepting a financially better offer for his inventions. This group was from Germany, associated with a major automobile corporation, and they apparently offered $100 million.

This time Jenkins accepted, but only if they met his conditions: "I get ten per cent of the gross for the rest of my life and in perpetuity, and I want it put into a trust. To do good stuff in the world. You have a five-year window to implement the technology, deploy it and show me that this is going out commercially—that it's not going to be buried."

The corporate reply was no. That was a deal breaker. He was told, "Just take your paycheck and go sit on the beach."

Jenkins replied that he already lives on the beach and doesn't need their money to be happy. He would rather that his inventions be useful to the people of Earth.

He has had moments in which he felt like kicking himself for rejecting offers, he told Jeane, then he reminds himself that the technology is meant to benefit people and not be buried by corporations.

HOLISTIC THINKING, FROM CHILDHOOD

Walt Jenkins' first role model of a self-motivated learner was his father, an electrical engineer in the U.S. Air Force who serviced radar stations and had been a bomber pilot in WW2. His father retired as a Colonel and, with Walt's mother, settled in Florida.

The family had lived in Japan in the 1950s when Walt was eight to twelve years old. The boy noticed that most Japanese people in that era seemed to think in a different way than Westerners. He also noticed temple paintings and wall carvings in Asia. Ancient cultures apparently had technologies such as the flying machines that religious texts in India called vimanas. Those cultures were far ahead of the West for millennia, he concluded. Regardless of wherever their advanced technology came from or how it disappeared, it sparked his imagination.

When the family moved back to America, he became aware of a contrast with Eastern thinking. As he matured, he viewed Western thinking as progressing faster technologically but being less holistic, not as burdened by considerations such as how an action affects the environment.

An innovator with a holistic worldview notices the problems facing humankind and holds the attitude "maybe I can fix them." For instance, years ago Jenkins had been reading about cargo ships powered by burning the crude, contaminated substance called bunker fuel. When it is burned, it pollutes heavily. Research shows that one giant container ship can emit nearly as much cancer-causing and asthma-causing chemicals as 50 million cars.[181]

Appalled by the problem of pollution from ships, he began years of experimentation using his process with sea water instead of tap water, and then pouring the processed water back into the sea. For the first few years of experiments, the end product was a grey slimy sludge which clogged up the device. Jenkins realized that the sludge would kill tiny forms of sea life, so he stopped and looked for approaches that respect living organisms. He began to work on a system that will return the rest of the water into the ocean in a condition as close to its original cleanliness as possible. He will release the sea water system for widespread use after he has proven sea life exits it unharmed.

The research had a beneficial byproduct, a way to take salt out of sea water that Jenkins says is cheaper than other desalination systems. Filing international patents on that is the next million-dollar step.

Professor Haralick was part of a group of independent scientists who visited Walt Jenkins in Florida in the spring of 2018 to view demonstrations. Dr. Haralick told the 2018 Energy Science and Technology Conference what he

---

181   https://www.theguardian.com/environment/2009/apr/09/shipping-pollution.

had seen during several different tests on engines using a mixture of mainly water as fuel. "So I think it's real."[182]

Later that summer, Walt Jenkins told us he had secured funding, but he didn't reveal the source.

---

182   Professor Haralick speculated that what is going on in the black box has to do with longitudinal waves (different than standard electrical transmission), among other things. Haralick said he has an idea of what is inside the black box because he had read Moray King's book.

CHAPTER 20
CONSERVATIVE
REVOLUTIONARIES

We need some unconventional thinkers—especially young,
brilliant, sharp-eyed thinkers—and we need to cheer not
sneer at their efforts.

—Huw Price, PhD, University of Cambridge professor.[183]

REPUTATIONS ARE A BIG CONCERN FOR SCIENTISTS WHO
need investors to fund their work.

Dr. Randell Mills has needed to distance himself from the field of research
that 30 years ago was misnamed cold fusion. Other scientists continued
in that field despite derision from nearly everyone else. However, most of
them, like Dr. Mills, distance themselves from anyone who has been labeled
"fringe," such as some of Nikola Tesla's fans. They also avoid controversial
concepts such as ether.

What we consider important, however, is that Dr. Mills' invention is a
well-documented breakthrough. And his saga touches on problems faced by
others whose inventions could revolutionize energy science and technology.

One validator described Dr. Mills' SunCell® as "a million watts in a tea
cup." It involves bursts of light and heat that are thousands of times more

---

183   Price, Huw, "Icebergs in the Room? Cold Fusion at Thirty," March 4, 2019, https://
www.3quarksdaily.com, Creative Commons article.

power-dense than any standard chemical process could release.[184] Credible scientists in various laboratories showed that Dr. Mills' invention releases massive amounts of non-polluting power from tiny amounts of water.

Dr. Mills' explanation of what happens in his device is controversial—hydrogen atoms in water are shocked into having a smaller orbit and therefore releasing energy. He named the shrunken atom *hydrino*. Dr. Mills has written more than a thousand pages of mathematics and other science to explain his "Grand Unified Theory." It is based on classical instead of quantum physics.[185]

Since "hydrinos" have unique properties, Dr. Mills was able to make new combinations of molecules. When he tried to patent his discoveries, however, one or more mainstream physicists interfered. After the U.S. Patent and Trademarks Office had already announced that a patent would be granted to Dr. Mills, certain physicists persuaded the officials to revoke the patent—on the grounds that his process is deemed unknown.

Dr. Paul LaViolette sums up the incident: "They made a big deal about it, complained to the patent office and claimed he was a quack. Because a few people didn't like his theory."[186]

For a quarter of a century Dr. Mills' work was mostly obscured by a wall of criticism, partly as a result of people who dismissed his work by erroneously categorizing it.

---

184 The BrilliantLight website says Each SunCell® has two electrodes that confine injected highly electrically-conductive, recycled inert molten silver metal and stable oxide that with supplied hydrogen from water as a source of reactants. A low-voltage, very high current ignites a reaction to form hydrinos and causes a burst of brilliant light-emitting plasma power of millions of watts that can be directly converted to electricity using concentrator photovoltaic conversion technology.

185 Mills, Randell L., *The Grand Unified Theory of Classical Physics,* Blacklight Power Inc. (2016).

186 Dr. LaViolette speaking to 2017 Energy Science and Technology Conference, Hayden, Idaho.

FLEISCHMANN-PONS HISTORY CAST A SHADOW

That type of dismissal had its roots in 1989, when two respected electrochemists, Drs. Martin Fleischmann and Stanley Pons, in Utah measured excess heat from an electrochemical experiment using the form of water called heavy water. Scientists assumed that Pons' and Fleischmann's excessive heat output was released by atoms that were fusing, but strangely the experiment did not require the intense heat and pressure that "hot fusion" requires, nor did it release dangerous particles. Instead, it was a tabletop operation and the electrochemists were unharmed.

Competition from an experimenter at another university caused the two men to be pressured into calling a press conference to announce their findings. The phrase "cold fusion" was then headlined around the world to describe their experiment. Drs. Fleischmann and Pons did not give a clear enough picture of exactly what caused the excess heat, and physicists generally did not have their electrochemistry skills. Therefore, teams at prestigious universities tried and failed to replicate the process. Universities were quick to dismiss it. The announcement of tabletop fusion had threatened the funding of expensive hot fusion research.

Science writers labeled the Fleischmann-Pons work "bad science," ridiculed them, and wrote obituaries for cold fusion. However, obituaries were premature and wrong. A colloquium held on the Massachusetts Institute of Technology campus in the spring of 2019 celebrated successes from 30 years of cold fusion research carried on despite shunning by mainstream science journals.

The cold fusion furor tainted the way people viewed Dr. Mills' invention, despite the fact that it is a different process and his findings are published in more than a hundred peer-reviewed science publications.

MILLS SEES FUTURE POWERED BY DROPS OF WATER

Mills began his line of discovery when he was in his twenties.[187] He named his company BlackLight Power, because his process radiates extreme ultra-violet high-energy light. A further step now converts the UV light to blindingly bright white light, inspiring the company's current name—Brilliant Light Power.

Dr. Mills once told Jeane that all of society's industries, homes and vehicles could be powered by tiny amounts of water, and he predicted his process would also work with sea water. He said the units will be compact like a small personal refrigerator. The size of a power source matters in homes, vehicles, boats and airplanes. Imagine an electrical substation compressed into a square meter of area.

He envisioned replacing everything that usually feeds an electrical substation. From high-voltage transformer and transmission lines to "the switch gear, the central power plant, the coal trains, strip mines, gas fields, oil fields, nuclear processing—everything's gone." They would be replaced at the local substation by megawatt units of his invention. A megawatt can power more than a thousand homes. It would create a decentralized mini-grid and much cheaper power.

Dr. Mills and his team built prototypes, one after another, because some of their ideas for harnessing electricity led to dead ends. At the time of this writing, the SunCell® is his best design.

What gave him the courage to start a hugely ambitious science-based company without government funding, without military contracts, and without the blessing of academia?

Randy Mills grew up on his family's grain farm in Pennsylvania—a good place to develop a work ethic and be resourceful and undaunted by setbacks. As with science, mechanics, chemistry and biology are an everyday part of

---

187   Ibid., Science Press, Ephrata PA 1996.

farm life.[188] He harvested hay and corn from his own leased acreage while still in high school. The successful young farmer had no plans to go to college.[189]

LIFE-THREATENING ACCIDENT

Mills' outlook abruptly changed after a life-threatening accident; he suffered severe blood loss from falling into a glass door. The trauma, and the five hours of surgery needed to repair his hand and arm, shook him. Thinking about an eventual death, he first wanted to know how everything—from his brain to the cosmos—works.

He enrolled in nearby Franklin & Marshall College, graduated at the head of his class, and went on to Harvard Medical school because he wanted to invent medical equipment. He began with pharmaceuticals and cancer therapy technology.

Mills completed his Harvard courses in three years, then a year of electrical engineering at Massachusetts Institute of Technology.[190] One MIT professor had insights about mathematics of a certain type of lasers and gave Mills a copy of the paper. Mills reasoned that he could apply similar mathematics to understanding the atom, which led him to solve longstanding physics problems.

After graduation he went home to Pennsylvania, partly because staying in an academic career could require a conformity of thinking he didn't want. He intended to work on his inventions, but medical products were soon sidelined. Instead, his excitement about mathematics of atoms led to his discovery involving hydrogen atoms.

While he ran electrolysis experiments in the kitchen of his apartment, he pored through books to find which catalyst (a substance that starts a chemical reaction and helps it to proceed) would fit his numbers. Electrolysis uses an electric current to drive a chemical reaction in a liquid. He published an

---

188   Jeane Manning's interview with Dr. Randell Mills, Feb. 10, 2000.

189   Baard, Erik, "The Quantum Leap," The Village Voice, published online Dec. 22-28, 1999.

190   Rosenblum, Art, "Randell Mills – New Energy and the Cosmic Hydrino Sea," *Infinite Energy* January 1998, p 32.

article and held a press conference in the county courthouse lobby in his town, Lancaster, Pennsylvania.

A local company, Thermacore, became involved. Their first experiment didn't work well, but they persevered and got power that was ten times more than the input.

One customer for Thermacore's industrial products was a German plasma physicist, professor Johannes Conrads. Through him, Dr. Mills met the head of the German physics society, who said he would take Mills seriously if he could solve wave-particle duality. That contradiction in quantum physics describes the electron as both a particle and a wave.

MILLS: '20TH CENTURY PHYSICS CHOSE WRONG DIRECTION'

The challenge led toward Dr. Mills' unified theory. Unified means it includes physics of both subatomic and the much larger scale of things. His theory did not need esoteric explanations such as ten-dimensional universes, wormholes connecting to other universes or virtual particles.

Professor Conrads didn't embrace the shrunken-atom theory, but his experiments produced remarkably high energy from a Mills cell. To Dr. Mills' delight, the German academic announced to the American Chemical Society that a catalytic reaction involving hydrogen produced power, a very high energy light, and was about ten times more energetic than any known chemical reaction.

Dr. Mills' power cells involved plasma. In a gas such as hydrogen, plasma is an area of the gas that conducts electricity easily. His experiments evolved to the point where the plasma could power the energy-yielding reaction on its own after the electrical heat source was turned off. That reaction lasted much longer than it would from any action recognized by standard physics.

Dr. Mills raised enough private capital to continue developing a product, so his company was able to hire specialists and buy a 53,000 square foot building near Princeton, New Jersey, to house its laboratory equipment. The skilled team eventually made high-energy bursts of light happen 1,000 times a second, resulting in continuous power density that the company's website says is 10,000 times brighter than sunlight.

To make an indestructible spherical dome to surround the reaction, they found a carbon material that survives high heat. The dome radiates illumination in the spectrum of light we can see. The dome in turn is surrounded by advanced photovoltaic cells that turn light into electricity.

The Brilliant Light Power company expects the electrical generation to cost less than 10 per cent of the cost of power from any known source. "The energy release of the hydrogen separated from H_2O, that can be acquired even from the humidity in the air, is over one hundred times that of an equivalent amount of high-octane gasoline."

LENR ALSO MAY END THE FOSSIL FUEL ERA

Companies that pursued other routes for getting power from small amounts of water have similar expectations. A research group named The New Fire posted a comparison of how much crude oil would be needed to match the energy output of a barrel of "LENR fuel" that could be a combination of half nickel and half two other elements. The group says it would take a supertanker load of crude oil—two million barrels—to match the energy output of one LENR barrel.

LENR is the most-used name for the mysterious heat-producing process, once called cold fusion, which perhaps is not even fusion of atoms. The letters stand for Low Energy Nuclear Reactions. Other rebrands of "cold fusion" include "condensed matter nuclear science" and LANR (lattice-assisted nuclear reactions). Those names were not the best public relations either. When we in the general public hear the word "nuclear," few people think, *Oh, the nuclei of atoms may be involved.* Instead, we may mistakenly associate the word with nuclear power plants that emit radioactivity, which is not the case with LENR processes.

Researchers at Google are exploring LENR, and so are the US Navy and other laboratories. And an earlier invention is still promoted as able to provide enough heat for industries—E-Cat, or Energy Catalyzer. Italian inventor

Andrea Rossi developed the E-Cat with help from physicist Sergio Focardi, PhD. Rossi keeps technical secrets private, yet his demonstrations impressed some scientists and attracted millions of dollars in funding.

A leading company in LENR is located in Berkeley, California. Scientists at Brillouin Energy claim to be developing a technology that will also be able to produce commercially useful amounts of heat. Excess heat from Brillouin's reactor has been proven at the Menlo Park, California, laboratory of Stanford Research Institute International.

As with Brilliant Light Power, the lead scientist started with a theory. In the case of Brillouin's Robert Godes, PhD, the theory allows certain metals to produce a "controlled electron capture reaction," on demand, using isotopes of hydrogen. The system is called a Hot Hydrogen Tube. Brillouin aims to power up to 30,000 homes for one year on the amount of hydrogen in an average glass of water.

The LENR field is populated by a high percentage of scientists with doctorate degrees and is well papered with more than 1,000 technical articles. *Infinite Energy* magazine chronicles the technical progress of LENR.[191] Ruby Carat's podcasts on ColdFusionNow.org are an entry point for the lay person who wants to hear about the science in that field.

LENR requires specialized knowledge and equipment and is not likely to be an answer anytime soon for the do-it-yourself maker community of people who want energy independence.

That ideal of energy freedom conflicts with corporate plans to have ubiquitous sensors harvesting citizens' behavioral data into digital information banks. With energy independence, you don't need a "smart" meter on your home. The rush to promote "smart" everything, from appliances to cities, is very visible. Are the sensors another push into an era of what Harvard Business School professor Shoshana Zuboff calls surveillance capitalism?[192]

Regarding the politics of energy, a physicist who was known in science magazines as Mr. Hydrogen Economy—professor John O'Malley Bockris, PhD—told Jeane that the public is more powerful than people realize, and could insist on new energy alternatives. "But the public is asleep."

---

191 https://lenr-canr.org.

192 Naughton, John, *The Guardian*, January 20, 2019. https://www.theguardian.com/technology.

That was decades ago. People are waking up. Apathy is yesterday's option. Activism is rising, especially among young people. Today, the mood of the public is changing due to fear of disasters from climate change.

PART IV
BODY, MIND & SPIRIT

CHAPTER 21
FOR LOVE OF THE LIFE FORCE

This is a voyage into a new era of science.

—Claude Swanson, PhD, physicist,
author of *Life Force: The Scientific Basis*

THE FIRST TIME PAUL BABCOCK AMBLED ONSTAGE AT THE
energy science conference in northern Idaho, he introduced himself in a
self-deprecating manner. The audience warmed to his humor. They saw a
tall man wearing blue jeans and a plaid shirt, sporting a ponytail and greying
beard and speaking unpretentiously and directly. Nevertheless, he exuded
dignity and a large presence, not only in physical size.

Babcock's casual manner was refreshing, coming from someone introduced
as an internationally-recognized technical expert.[193] He spoke with conviction
as well. "At this time in our civilization we have to decentralize power. All I
want to do is give you people the tools to advance the cause."

He explained which electrical laws are never subject to change and which
ones could be skirted in order to engineer electrical circuits that harness
magnetism. Simple algebraic truths prove that magnetism imparts energy,
he said.[194] Amplification of power is how the universe works and there are
ways to tap magnetism.

---

193 The Bedini-Lindemann conference 2012, beginning annual series, now known as the
Energy Science and Technology Conference.
194 The physics laws named after people such as Ohm, Faraday, Joule, Lenz and Kirchhoff
are essential, Babcock says, for anyone who wants to do free-energy work out in the world.

Babcock was asked when his company[195] will develop the prototype of his energy invention into a commercial motor. Not until he can pay the million-dollar cost of product development himself, he replied vehemently.[196] Not until he doesn't have to beg for start-up money and go to financial institutions for funding.

To raise the money needed to optimize the motor, his company Flyback Energy was making ultra-efficient industrial lighting products, expecting that they would be more acceptable to vested interests than free energy.

It didn't turn out to be that easy.

THE MAKING OF AN INVENTOR

The struggle to make a revolutionary energy breakthrough available to the human family requires strengths. Paul Babcock's embattled yet often joyous boyhood prepared him.

Growing up in the rural center of Washington state in the 1960s meant freedom to roam outdoors and have access to the natural world—a good start for a scientifically inquiring mind. He was in an inventive family; an uncle invented farm machinery. Paul and his brother Dave were "electron-chasers" from childhood on, experimenting with electricity.

They had a pack of lively siblings and friends to run with and an abundance of both comedic and ethical examples at home. Their parents encouraged intellectual pursuits as well as insisting on values such as honesty.

The hurtful battles took place outside the home. The small town of Omak, Washington, prides itself on a Wild West heritage. But unlike cowboys-and-Indians movies with heroic good guys, real life in what he called a reservation town revealed ruder traits. The seven Babcock brothers and their two sisters encountered bullies and small-minded attitudes. Their mother and her family were tribal members and were targeted by local racists. The Babcock brothers learned how to fight.

---

195   https://www.flybackenergy.com.
196   Babcock, Paul, public speech in 2012.

Those blows toughened Paul Babcock. As it turned out, he needed a thick skin if he was going to promote inventions that disrupt established interest groups. He also questioned authority from a young age.[197]

THE JOY OF DISCOVERY

Babcock thanks his parents for his boyhood freedom to experience everything from hefty power tools to dangerous science experiments. He enjoyed "dogs, cats, terrariums, aquariums, snakes, lizards, fish, bugs, radios, electronics, model airplanes, minibikes, mountaineering, camping, hunting, fishing, river running, sandlot sports, cars and more...always a lot of action going on at the Babcocks."

Craving knowledge, he read avidly. The town library's staff noticed what a range of science books young Paul Babcock borrowed, so they introduced him to Dr. Wilhelm Reich's science based on a universal life force that Reich named orgone.[198] Public libraries at that time were pressured, apparently by a government agency, to pull books written by Dr. Reich off their shelves. The federal Food and Drug Administration had publicly burned Reich's books. The Omak librarians quietly kept some. As a result, young Paul learned that scientific experiments prove the existence of that vital life force.

OUT IN THE WORLD

Babcock moved to Arizona and learned classical electrical physics in college. However, during his more than 30-year career in the industrial world he and his brother Dave saw phenomena that defied conventional understanding. While working with high-powered electrical systems in remote locations as a trouble-shooter, he would suddenly see bursts of electricity seemingly come out of nowhere.

---

197 Paul Babcock interviewed by Jeane Manning, summer 2016.

198 Well documented film on life of Wilhelm Reich: http://loveworkknowledge.com.

Frustrated by the fact that the bursts didn't relate to the standard Ohm's law of electricity, he studied technical writings of Nikola Tesla and others such as inventor John Bedini. Something was missing from standard physics teachings.

EXPLORING, WITH AWE AND WONDER

Babcock's life laid the groundwork for his conclusion that what is missing is the living process referred to as Mother Nature. "I love being a tech guy," he said. "At the same time, I had to grab my backpack and head into the woods for weeks at a time. So, I lived in the Alaska bush for years… I've damn near died frozen on mountain tops, or in deserts for lack of water…. The 'invisible hand' has reached out and saved my life more times than I can count."[199]

His experiences ranged from swimming for miles with dolphins and being protected by them to being humbled by a caribou that stood still and gave its life so he could survive a winter. He felt gratitude for fish and seaweed that kept him alive on a rocky island near Hawaii, and for seeing the hues of the life force. In Hawaii, more than one Kahuna gifted him with knowledge about the life force we inhale with every breath. In settings as varied as a prayer circle on a reservation or a whale hunt in the arctic, he learned from indigenous peoples about hidden energies.

Experiences such as seeing auras he mainly kept to himself; they weren't topics to share on a job site. You don't tell the guy on the barstool beside you that water is a conscious aware entity, Babcock realized. However, he was certain that a person's inner senses are as valid as their physical senses of sight, sound, touch, smell and taste.

---

199 Babcock, Paul, *The Living Earth*, 2018 Energy Science and Technology Conference.

A NEW LOOK AT MAGNETISM

Babcock went north to Alaska in 1978, worked in the aviation industry, then booked a year off for wilderness adventure. After that he went into radio communications. Job sites ranged from oil wells to mountaintop wind generators.

His quest for answers about energy intensified in 1988 in Alaska. During a run of bad luck, he was unemployed because his right arm had been injured in an accident and was in a cast. During a severe winter he survived with little money or food.

In his small cabin lit by a kerosene lamp, one night he heard a radio show whose guest was an inventor named Joseph Newman, talking about a "free-energy machine." Newman's claims about magnetism intrigued Babcock, so he scraped up fifty dollars and ordered Newman's book.

"I got this big tome, a huge book...politics and ego and patent wars," Babcock recalls. "But... Newman made this simple observation about magnetism: 'the power you expend to make a magnetic field has nothing to do with the strength of the magnetic field you create!'" Babcock set out to learn how Newman's finding relates to the known laws of electrical physics and what that would mean for humankind.[200]

An "aha!" moment came when he was still in Alaska. He was sitting by the wood-burning stove at a friend's house when he encountered new very powerful magnets made from an alloy of the element neodymium. The friend handed Babcock two magnets, about an inch and a half square each, and the men marveled at the super-strong forces of attraction and repulsion.

When Babcock reached out to give one back, his hand passed near the cast iron stove. The magnet suddenly leapt off his hand and accelerated. It hit the stove with such force that the magnet shattered and the metal stove rang like a bell.

Babcock realized that the acceleration—a huge expression of energy—had nothing to do with his hand's motion which was relatively slow. Instead the magnet imparted the energy itself. He hadn't lit a rocket or burned any fuel to make the magnet speed up so forcefully.

---

200   Geoffrey Miller replicates and explains the late Joseph Newman's machines; see https://energybat.com.

"Wow. That was an expression of work, of horsepower; it moved mass over a distance with foot-pounds of force!"

That incident fueled his quest to tap magnetism for a power gain in electrical circuits. First, he had to create an electronic circuit that can handle exceptionally strong magnetism. And it would require extremely fast electronic switching. How could he accomplish nano-second switching while living in Alaskan bush country with only occasional consulting jobs? The state-of-the-art electronic parts he needed were too expensive.

His brother Dave and their friend Phil Smith had the answer. They were working for a Seattle company that regularly discarded electronic parts the company wasn't selling. Even if needing repair, the equipment his brothers shipped to Alaska was like gold to resourceful specialist Paul Babcock.

The brothers and Smith increasingly focused on evolving the circuit which would become a key part of the Babcock DC Motor and other products. Paul moved back to Washington state to rejoin his profession long enough to update on the latest in electronics.

Creating the circuit and motor involved further "aha" moments, perseverance and financial sacrifices from each partner. Paul Babcock jokes about them putting one prototype together with masking tape and bubble gum. High hopes often plummeted. Promised funding disappeared after the international trauma of September 11, 2001.

He emphasizes that the team never claimed to have a free energy machine, but rather a motor that is super-efficient because it captures wasted power and recycles electricity.

"WHAT PART OF SEEING IT DON'T YOU UNDERSTAND?"

Babcock and his partners went on the road years later, expecting that experts would welcome revolutionary Babcock DC Motor. The crew could show a prototype in action, and they had professional accomplishments in the electrical industry as credentials. Further, Paul Babcock could explain in standard technical language why the motor can do what it does, despite textbook laws that say such performance is impossible.

The team didn't get a chance to explain it at universities where the motor could have been tested. As soon as the professors realized that a demonstration could contradict accepted theories about electricity, they responded to the effect of "Get out of here, you crazy hippies!" and more restrained variations of hustling the Babcock crew out the door.

The rejections dimmed the likelihood of speedily engineering the prototype into a product you can buy. Commercial development takes millions of dollars, so an inventor must attract investors, but investors want to see signed papers showing "expert validation" for the invention. The "money guy" walks away if an inventor can't get endorsements from professors at a university. However, the potential investor often does not realize that the professors are experts in a different paradigm.

INDUSTRIAL LIGHTING PRODUCT

By the time that Babcock was invited to present to a more open-minded audience, the 2012 Energy Science and Technology Conference in Hayden, Idaho, his motor was gathering dust in a home workshop. He was spending long days in a high-ceilinged warehouse in Spokane Valley where he and his brothers and Phil Smith developed an industrial lighting product that incorporated their revolutionary electronic circuitry. They hoped the product would soon bring in money for further development of the original motor.

They named their company Flyback Energy. Flyback is the burst of energy that suddenly appears when an electrical current's flow through a conductor is interrupted and as a result the magnetic field that had been produced by current flow collapses, creating an electrical discharge. Although physicists have believed the electrical kick is not energy, the company turned the problematic kick into a solution for their business of *magnetic energy recovery*.

Their company's struggles reveal how much is involved in bringing even a niche product such as ultra-efficient ballast for industrial lighting into the marketplace. It costs millions of dollars to get patent protection and

Underwriters Laboratory certification. Each round of fund raising cost the partners in different ways, such as losing some control of their company. And the company holds the patents.

Babcock no longer talks publicly about the Flyback history except to say, "No one can screw up a project like the guy with the checkbook can."

COLLABORATION

Following his first presentation to the energy technology conferences in Hayden, Idaho, Babcock was invited back as a speaker each summer. The conference venue is a fraternal lodge in a semi-rural area, yet engineers from far-flung countries meet across tables in the lunchroom or bar between speeches in the main hall.

That's where Babcock and Jim Murray (Chapter 10) met and discovered a synchronicity. Babcock's toroidal motor sitting in his workshop was feature-for-feature almost identical in concept to a toroid-shaped motor back in Oklahoma that Murray had independently invented. The two inventors joined forces for a project—a few months of intense work adding Babcock's nanoseconds-switching breakthrough to Murray's Switched Energy Resonant Power Supply. They also succeeded in working out the mathematics.

Babcock notes how Murray's work relates to past and present discoveries. "Nikola Tesla used high voltage and particular forms of resonance to confuse and manipulate electron motion for the purpose of tapping 'Radiant Energy,' in modern times known as zero-point energy."[201]

HEALING

Babcock's life took a turn into alternative research for health. His colleague Dr. Peter Lindemann built a device whose inventor[202] had named the

---

201 http://paulmariobabcock.com/blog.
202 Belarusian-French engineer Georges Lakhovsky developed the original Multiwave Oscillator from the 1920's to the 1940's.

Multiwave Oscillator. The MWO was based on the principle that life-forms absorb energy. Babcock built a powerful MWO and presented his findings at the Energy Science and Technology Conference in Idaho.

When Babcock took the stage at the 2018 conference, he surprised those who only knew of him as a seasoned technician or an inventor who created products based on recovery of magnetic energy.

His presentation *The Living Earth* went beyond technology, starting with the need to balance logic with intuition. "Science based solely on hard reason and the five senses…harmed mankind more than it helped…Denial that we're a spirit being as well as a physical being degrades the sacredness of life." Babcock also said his life had taught him that every soul is here in their physical life to play as well as learn. "We are conscious creators with the Big Holy…it's utterly fun."

The audience applauded his admission that, "I love free energy; it's very necessary for us, but if we don't grow as human beings, if we don't find who we really are, if we don't master ourselves better, what's the point? One day it dawned on me…'If I give the world this stuff and all it does is make cheaper electricity so people can watch more TV, what have I really done?'"

The buoyant mood in the room sobered as Babcock warned about metal particles found in the fallout from airplanes' aerosol droppings. He said barium and aluminum particles create electrically-conductive paths that degrade the atmosphere's ability to electrically support life.[203] He explained that fluffy cumulus clouds, which children instinctively love to draw, interact with natural lines of force as they move along. That electrical process replenishes, and water stores, the life force energy, he said. Healthy clouds can carry tons of water held together by static electricity forces.

He found that solar flares and weather events affect how his Multiwave Oscillator behaves.[204] A properly grounded MWO with a field of more than 100,000 volts interacts with the atmosphere. He concluded that coordinated use of MWO devices operating at Earth friendly frequencies could restore the local dielectric field in the atmosphere by interrupting the unnatural activities.

203   https://emediapress.com/2018/08/30/the-living-earth-how-earth-generates-life-force-energy-from-the-radiant-energy-medium-by-paul-babcock.
204   A Multi-Wave Oscillator is a therapeutic device.

However, there is a better way to deal with the human-caused problem than technological solutions, Babcock said. "The real solution is for human beings to embrace their full talents as reality creators and take conscious control of our relationship with the living vibrating medium."

CHAPTER 22
WAYS OF KNOWING

Only biological intelligence can make machines self-run;
no matter how much you spend on software, artificial
intelligence doesn't know how to work with the aether.

—Alan Francoeur, inventor.

NEW ENERGY INNOVATORS GET BREAKTHROUGH INSIGHTS
in a variety of ways. Alan Francoeur is a student of nature's motions. He found
that knowledge of how to make a better power generator can be hidden in
the natural world and revealed when a person persists.

He was a teenager when he became a mill operator in Canada's Northwest
Territories at Yellowknife Giant Mines. The mine's immense motors and
generators fascinated him and he analyzed them so often that he found
himself daydreaming about magnetic fields in motion. He had a vision of the
fields flowing like water and wondered how their motion could be enhanced.

After work hours he tinkered with magnets, a secondhand voltmeter and
makeshift electrical coils. He cut off pieces of tin to insert between and
quickly remove from stationary coils and magnets. His curiosity about the
effects of that shielding interference on magnetic fields grew into an obsession.
When he moved to Calgary a few years later, he started building what he
called an Interference Disk Generator, based on what he had learned. Other
experiments created fuel-saving inventions at the same time as he became
a single parent. There had been no father presence in his life, so when he

took on the responsibility of parenting his son he again figured out how to deal with new challenges.

A Calgary doctor decided to pay the 20-year-old inventor's travel expenses to the Gravity-Field Energy Congress, in Hannover, Germany. A decade or more later, Jeane met up with him again. By then he was blessed with a daughter as well as his teenaged son and a cheerful, supportive wife, Jan. They lived in Penticton, a small city in British Columbia. His projects, tools, bookcases and boxes of machine parts filled their small living room.

Expensive magnets, electronic parts and machine shop bills are the price of building a novel generator, and Francoeur is fortunate to have a wife who agreed to the financial sacrifices. Jan has been by his side when new insights burst into his mind, such as an understanding about electromagnetic standing waves in certain resonant electrical circuits.

That insight came when they stood on a lakeshore observing a standing wave created when the wake behind one speedboat merged with another boat's wake in a particular way. At other times he could sense a life force in the atmosphere, a phenomenon not acknowledged in standard science writings.

The couple's ethics were tested in decisions that at first appeared as a financial opportunity.

Francoeur hosted an online forum and shared his technical ideas. A wealthy American to whom we give the pseudonym Mr. Rufus Pemberton III became interested in Francoeur's innovations and flew to Penticton in a private aircraft. Pemberton made it clear that he was "old money" and therefore entitled to special treatment.

Francoeur designed industrial apparatus that would solve expensive problems for the mining operation Pemberton planned to buy. Francoeur's intentions were both to save money for the industrialist and prevent environmental pollution. Meanwhile the Francoeurs had doubts about the rightness of working for Pemberton. That intensified when the industrialist put a condition on their proposed contract; he expected Francoeur to be an employee without autonomy over personal decisions such as his own diet. The clincher that made the Francoeur household firmly close the door on the deal was seeing Pemberton's rudely racist reaction to a person of color.

"You are not going to get the technology" was Francoeur's message to Pemberton. Francoeur later told us that he wants humankind to profit from his inventions, not an investor who demonstrates a low level of character or integrity.

The morally high road is not Easy Street, and the Francoeurs took whatever jobs were available. His long stint of building mobile homes for a manufacturer left him with chronic wrist and hand pain for years.

His enthusiasm for life was again on the upswing when he and his wife Jan found an opportunity to homestead on a forested mountainside above a tiny community in British Columbia. He soon had a workshop, in a shed close to their home.

From the Penticton days onward, Francoeur had studied information about motors built by an American inventor, the late Edwin V. Gray (1925-1989). Gray named his motor the EMA. It could produce 80 horsepower of work from four deep-cycle batteries that the EMA system recharged at the same time. "The claims were outrageous," science historian Dr. Peter Lindemann wrote about Gray's EMA, "but were supported by much of the testing data."

Gray's motors are said to have involved unusual 'cold electricity' *and* normal positive electrical energy. Dr. Lindemann's website has an archival photo of children plunging their hands into a bowl of water in which an electric lightbulb is radiating light.[205]

Jeane once witnessed a similar sight when an experimenter had followed Nikola Tesla's patents precisely to build what he called the hairpin circuit. It produced an unusual safe form of electricity. A light bulb lit by that was plunged into a bowl of water and audience members touched it safely. (Don't try this at home!)

When Francoeur learned that a collector was selling Gray's original machines, Jan supported his decision to buy the vintage motors despite their own financial needs. For instance, Alan had gone without dental

Alan Francoeur, inventor, in his mountainside workshop
(Jeane Manning photo)

---

205   http://free-energy.ws/edwin-gray.

care and lost his front teeth but postponed replacing them and instead bought magnets for experiments.

Francoeur took the vintage motors apart carefully and cleaned each part or found the same model to replace it. He explains that a type of extrasensory perception called psychometry seemed be giving him subtle information about each object as he reverently touched it. Everyone has the potential for accessing information from a non-physical source, he said. His experiences range from the common "tingling at the back of my neck" that alerts him to be aware, to what he calls a dump of technical information. At one point, he and his family saw unexplained aerial phenomena in the night sky over their mountainside homestead. The spectacle lasted so long that they were able to snap photographs. The pictures showed specific patterns of loops of light that seemed to him to be technical instructions for what he had been building on his workbench.

Francoeur says engineers could be building more efficient generators if they heed his insight that "mirror image" electrical windings are more harmonious with the spiraling flows of energy than standard electrical coils, and could be ultra-efficient. He sees examples in nature where mirrored symmetry appears, at scales from the micro level up to galactic.

When invited to explain to a 2017 energy conference how he believes the Ed Gray device's power supply works, he shared openly.[206] Afterward an international business owner who attended the conference offered him monetary help. Francoeur asked what the philanthropist expected in return and was told, "I don't want anything out of it." The grateful inventor in turn asked the conference organizer, who sells DVDs of the speakers' presentations, to donate royalties from Francoeur's presentation to a local charity that helps the homeless.

Francoeur remains adamant that nature uses "helical geometry and mirror-image symmetry" principles. Technology should reflect that, but he says electrical coils are usually wound in a way that opposes natural flows of electricity. His solution is helical windings incorporating that principle.

His ideas are sparking interest. The Francoeurs were invited to a conference in Shanghai, China, in August 2018, where he spoke to an audience that

---

206   Francoeur, Alan, "Electricity, Gravity, Magnetism & Singularity," http://emediapress. com/alfrancoeur/electricity.

included academic scientists from various countries. In 2019 his company EMA Magnetic Energy Systems, Inc. received private funding intended to develop Francoeur's inventions and restore Gray's machine and make it run in resonance mode. He describes the EMA technologies as "standing wave pulse motors supporting the electric universe model of a helix, ushering in an era of technological revolutionary change to benefit all humanity."

BOOK LANDS IN TEWARI'S HANDS

Inventors also get solutions to technical issues in dreams or other ways.

When Jeane spoke with Paramahamsa Tewari on a railroad platform after meetings in Colorado, he was certain of having received help from a beyond-the-physical source. For instance, while his demanding day job of building a nuclear power plant in India took so much time, when he did get into a technical library he didn't have hours to spend seeking information. Almost miraculously, one time the right book that he needed literally fell off the library shelf into his hands.

GRANDER'S INNER VOICE

Johann Grander of Austria did not have the formal education of a professional engineer such as Tewari. However, Grander found that the clean environment in the Alps mountains enhanced his thinking. His associates said that when Grander's motor project stalled he would go sit on his porch in the sun. He told them later that after about ten minutes he would hear an inner voice telling him to build the motor in another way.

His team likened it to the story of Johannes Brahms. When asked how he composed his famous music, Brahms had said that when he was in harmony with nature the music would come—not step by step but all at once and complete.

MERKL, SOUL TRAVELER

The late George Merkl experienced astonishing ways to learn about everything ranging from minerals to outer space. As a result he patented dozens of inventions. In formal schooling he earned more than one PhD degree, but after disillusionment with academia he discouraged people from calling him Dr. Merkl.

A biologist friend and Jeane interviewed George Merkl at his home in the hills above El Paso, Texas, in the year following the announcement of tabletop "cold fusion." Dr. Merkl said he had already discovered a similar process but had been harassed for announcing a disruptive energy breakthrough on national television.[207] He had then turned his attention to experiments with living plants and health-enhancing products based on his advanced knowledge of the universal life force, which he called "scroll energy" because of its spiraling movements.

The experiences that gave Merkl such insights began when he was five years old and living with grandparents. On a scheduled day for his mother to visit, the streetcar arrived but she wasn't on it. As the streetcar left, the frustrated child grabbed onto its rear and was thrown onto the cobblestones head first. Little George was in a coma for days. He reported that he journeyed out-of-body during that time. Afterward, the child would sit on the steps outside the house in the early mornings and wait for the sun to appear, as if the sun had become a beloved friend.

When George Merkl was an adult and imprisoned in his family's Budapest home during wartime, he passed much time productively doing lucid 'soul travel.' For instance, he said he learned more about mineralogy on distant planets while out of his physical body than he ever learned in a classroom

In his El Paso home many years later, Dr. Merkl and his wife Maria lived quietly among esthetically pleasing surroundings. Visitors felt they were in the presence of an extraordinary soul; the scientist's eyes seemed to radiate a gentle light.

---

207   Merkl, George, United States Patent 3993595
"Activated aluminum and method of preparation thereof," 1976.Nerjk

His conversation was a challenge to follow because he sometimes used his own poetic terminology. He showed paintings he had done to express advanced scientific concepts visually; an animating life force was a central idea.

Dr. Merkl's writings had common elements with what Dr. LaViolette and other pioneers of new science are saying.

CHAPTER 23
SHAPING LIGHT AND SOUND

Physics is on an amazing rollercoaster; it is set to change our world dramatically within five years.

—Anna Hall, materials scientist, mathematician and musician.

NIKOLA TESLA'S WORDS "IF YOU WANT TO FIND THE secrets of the universe, think in terms of energy, frequency and vibration" place discoveries of Randolph W. Masters, PhD, in a context. Dr. Masters' musical mathematics of light and sound could be a foundation for future science.

A way to *see* sound waves was first popularized by German physicist Ernst Chladni. He sprinkled sand on a flat plate, vibrated it by pulling a violin bow across the plate's edge, and saw geometrical patterns in the sand. Hans Jenny of Switzerland expanded those images into complex three-dimensional forms.

Dr. Masters studied the Great Pyramid in Egypt; it embodies concepts involving vibrations.[208] Its measurements include the 'golden section'—a number found in spiral ratios in leaves growing on a tree, whorls on flowers, spiraling vortexes forming toroidal blueprints in pine cones and seashells and elsewhere within nature's infinite fractal beauty.

---

208   For example, the mathematical factor pi is used to calculate the slope of a side of the Great Pyramid and is also used in electronics when relating frequencies.

Mathematical clues and other findings of Dr. Masters and of many researchers[209] imply that ancient pyramid builders were scientifically advanced. They also imply that the basic invisible substrate out of which the natural world is created flows into or resonates with certain geometrical shapes more than with others.

A SCIENCE LEADING TO AWE AND WONDER

Dr. Masters emphasizes the key role of consciousness in creating technologies that are in harmony with nature's ways. He intends to establish "conscious harmonic universe sciences" as a valid scientific pursuit, teaching harmonic laws that transcend music and including precise harmonics of the universal information field, the life field. Mainstream physics recognizes just four forces, yet research in the new branch of science points to a fifth force.

MAESTRO OF THE SINGING UNIVERSE

Dr. Masters is well named; he mastered multiple arts, sciences and musical instruments. He teaches in the Sound and Consciousness program at Globe Institute in San Francisco and at a community school of music and arts. His students have won prestigious awards. Meanwhile he earned a PhD in Divinity as well as a Bachelor of Arts in film and music.

He didn't expect to be teaching the physics of music, but he uncovered knowledge that has been "…edited out of books." When he began to see that, he was enjoying his career of teaching at universities of California, Santa Cruz, and San Jose State, performing with world-renowned musicians and winning awards. Today, his accolades include being called the Maestro of the Singing Universe.[210]

---

209   Books by Graham Hancock, Christopher Dunn, Robert M. Schoch, Robert Bauval, and articles in *Atlantis Rising* magazine, for instance, take a fresh and well researched look at ancient sites.

210   http://www.universalsong.net/about_randy.htm.

His mission, as he sees it, is to bring hidden knowledge back into proper use for tuning consciousness. He aims to bridge science and spirituality. "It's really *spiritual mathematics*, understanding nature and Creation."

It turns out that he could not have found his soul mission without first learning music theory with acoustics, mathematics, physics and harmonic structures so thoroughly that he saw everything from a musical point of view. Steeped in that analysis and in his private study of metaphysics, he examined archeology of ancient sites in extreme detail.

MUSIC HISTORY SHEDS LIGHT

He had learned musical ratios and rhythmic patterns used by different cultures and how much of it is built on whole numbers. Increasingly, he saw those ratios at ancient sites all over the world. Certain secret or sacred numbers were built into the sites because those numbers resonate with the universal energy field, he concluded.

Dr. Masters is writing a book for people knowledgeable in mathematics and physics. He will also provide books for lay people. His work points toward the more harmonious civilization that is possible on Earth. For instance, schools of higher learning could teach how to create what he calls "livingness technologies." Many researchers are calling it Nature Tech.

The sciences today overlook higher laws that govern music and life and are intrinsic to the way the natural world creates, he said. Much mathematics is what he calls "cerebral head-tripping"—not necessarily based on actual observations in nature or describing how creation works. The original foundation for understanding the universe was called the Musica Universalis, and Dr. Masters' research is enabled by seeing music throughout everything across the universe.

SACRED GEOMETRY

The popularity of books and films such as The DaVinci Code is due to humankind's awakening to the "music of the spheres," Dr. Masters noted. Those books allude to sacred geometry, which means the harmonic proportions found throughout the universe. The proportions are an essential element in nature's blueprints. Inventors of breakthrough energy devices sometimes set out to include such geometry when designing their devices.

The word geometry started out meaning to measure the Earth, Dr. Masters said. When we cite "sacred geometry," we are saying that builders use nature's favorite ratios and proportions in design, layout, and construction, whether of a temple, monument, church, aqueduct, or house. Thus, he added, every such construction is a prayer of gratitude and praise to creation.

During his spiritual studies, Dr. Masters had learned about the *audible sound current* as a divine creative force. His work today graphs out aspects of that sound current "for applications affecting consciousness through different frequencies of light and sound... and ratios and shapes... The sacred geometry is basically shape *resonance*. That's the key word."

KNOWLEDGE FOR HEALING

Things are the shape they are for various reasons, sometimes because of how they impact sound and how sound impacts their shape. That's why languages have expressions like "bent out of shape" or "we're out of shape," Dr. Masters said. For instance, when a kidney gets too far out of shape it doesn't function correctly even though it's made out of kidney material. It happens to be a key organ for sound, so traditional Chinese medicine practitioners clear out the kidney meridian of the body's subtle energy currents when a patient has a hearing problem.

Dr. Masters sees the information as helpful holistically. For instance, clearing a problem with hearing could allow a person to begin to hear the consciousness that's been avoided or obscured.

DESIGNING USEFUL THINGS

His private consulting work includes designing product containers for industrial clients by putting precise numbers into design factors. Proportions aren't enough, he said.

Of all the number versions that nature uses, certain combinations are more favorable than others because they act like a lock that must be opened with more than one person's keys. His analogy describes a combination so complex that key-keepers must coordinate before the lock can open. He has to find "a very specific wavelength, because that wavelength talks to reality like a special cellphone...This artifact has to be tuned using the numbers that talk to the universe's information field...with the correct wavelengths *and* ratio."

ANGLES AND ANGELS

Whether he is designing jewelry or a container for a health product, he has to know both the exact wavelengths emitted by the material and the correct shape to tune into the cosmos. Finding the exact angles to build into an object is also about harmonically bending light. "Angels and angles... are bringers of information."

Dr. Masters points out further insights built into human language. People say they see "a different angle on it." An animal, as well, will change the angle of its head to receive from the cosmos differently compared to the previous head positions, he said.

To amplify the intent of the contents of a container, he doesn't have to change the contents. Instead he changes the resonance of the container. "The numbers talk to the universe's information field."

WATER

If someone wants to invent a product for stirring liquids and hydrating, for example, design factors would use hydration frequencies as wavelengths. "So when you put that wand, say, in water and stir or set it on the top, the harmonics... are completely in line with the desired effects of penetrating deeply into the consciousness of water, both at the three-dimensional level and the omni-dimensional level of water."

Water is constantly communicating between dimensions, Dr. Masters said, and most scientists haven't figured out how magical water is because they're looking at it as only three-dimensional. Water is "going back and forth between the dodecahedron and icosohedron, shape-shifting the hydrogen bond back and forth. It's created an antenna for talking with the information field."[211]

The information field may be what is historically called ether. Dr. Master thinks this ether or a fifth force is responsible for and links the other forces—gravity, electromagnetism, and what physicists call the strong and weak forces. "It's like a universe level conductor creating a conscious harmonic universe symphony that delivers coherence to everything. It's the very essence of what the Chinese referred to in the I-Ching, explaining that the yin and yang of everything is balanced."

As a public speaker and a teacher of Universal Song classes, Dr. Masters helps people "connect the dots of a far more loving and intelligent universe, where consciousness is seen in creation's harmony between all things, leading to more respect for our existence and all levels throughout the cosmos."

For example, he said, "if you could clean up all the water in the world you would clear the path for us to feel much better emotionally and physically. Pollution in water creates a different field as well as effecting our biology. Clean it up and you're actually cleaning up consciousness."

Enough water to fill up water containers can be drawn out of the air in a desert, Dr. Masters said. It requires knowing how to properly "work with the field of water" with ancient technologies.

---

211   Different platonic solids relate to different elements, in Dr. Randy Masters' view; the dodedcahedron bridges between water and the life force, the ether field, while the icosahedron represents water.

YIN AND YANG OF KNOWLEDGE

Dr. Masters talks about feminine knowledge as being the night side of nature and analogous to undertones in music. Overtones are multiples of a frequency and audible to us, while undertones are divisions. Even if you don't hear an undertone with your ears, it supports the music we do hear.

Discussions about the yin and yang (Chapter 26) can get mired in duality. Dr. Masters advises that in order to create devices that access the ether, researchers need to recognize a "neutral still point" aspect. Properly balanced without conflicting agendas, male and female energies create the neutral force. The result is a trinity of wholeness. Balance would *not* mean a neutral point dividing in the middle like a playground teeter-totter. To be harmonically creative, a different proportion is needed for the female component and the male.

Such principles and design factors may create useful things that enhance life instead of destroying.

Nearly every device Dr. Masters sees could benefit if "conscious harmonic universe" thinking was applied to it. Scientists have powerful tools yet might not have spiritual maturity to even guess at the possibility of accidentally exceeding the current known laws of nature, he said. "So much of nature and its harmonious laws are yet to be discovered, it's almost as if they are veiled until our global wisdom reaches the right maturity."

Currently there is no international wisdom council investigating new inventions, gathering information on possible long-term effects of a technology, and overseeing the use of advanced science. Dr. Masters challenges humankind to respect all living beings including Mother Earth. He believes our success as a species relies on working together for the good of all, and that decision-makers should look at all technologies in terms of their 'livingness" and conscious harmony with the cosmos.

CHAPTER 24
ACADEMICS, ASTRONAUTS
AND CONSCIOUSNESS

To understand the things that are at our door is the best
preparation for understanding those that lie beyond.

—Hypatia of Alexandria, 350-415 AD

SUSAN'S MENTORS HAVE INCLUDED FORMER ASTRONAUT Edgar Mitchell, PhD, as well as Harvard professor of psychiatry John E. Mack, M.D. A colleague of Dr. Mitchell and Dr. Mack who also influenced her is astrophysicist Rudolph E. Schild, PhD. His own insights are furthering the former astronaut's theory about the universe and consciousness.

Briefly, the psychiatrist researched human experiences of being contacted by other intelligences. The astronaut and the astrophysicist developed theories about how the universe makes such experiences possible. In other words, Dr. Mitchell developed and Dr. Schild further describes a model of the universe that could explain unusual ways that individuals throughout history have gained knowledge.

Why should we care that these scientists are using accepted mathematics to underpin their hypotheses about consciousness?

Susan believes their work does relate to the New Energy field. "If human beings are indeed connected by frequencies with the physical makeup of the universe, then some of our most creative inventors—those working on technologies in alignment with nature's energy—may themselves be aligned to this frequency."

And the work of Drs. Mitchell and Schild sheds light on experiences we've heard about from inventors of energy breakthroughs, so we'll look at their journeys of discovery.

AWE AND WONDER IN OUTER SPACE

People watched their television screens in amazement during the February, 1971, Apollo 14 voyage to the moon. Susan is of a younger generation and wasn't part of that audience, but on a trip back from Rangeley, Maine, in the fall of 2007 she heard firsthand from Dr. Mitchell (1930-2016) about his journey. He described his experience of returning to Earth from the Moon as being in a 'samadhi state' in which his mind was still and his heart felt oneness with all living beings.

Imagine being one of the first to see Earth as a small, majestic planet in blue and white splendor floating in a dark sky. Dr. Mitchell said that sight stayed with him profoundly.[212] When he gazed at our planet from a quarter of a million miles away, he imagined what Earth can be if humankind chooses to make everything work in accordance with the natural design of the universe.

He could see that a global change of consciousness is needed. The universe does not bow to humans. Instead, we who inhabit "an insignificant little planet" must find ways to bring our consciousness into attunement with the universe, Dr. Mitchell had said.[213]

Those perceptions stayed with him, long after the splashdown and the heroes' welcome and parades were forgotten. After leaving NASA, he founded the Institute of Noetic Sciences which is devoted to exploring the power of the mind.

Dr. Mitchell's Quantum Hologram Theory of Consciousness sheds light on the nature of non-ordinary states of consciousness and how knowledge

212   Ibid.
213   Capt. Edgar D. Mitchell PhD, *"A Change of Consciousness"* foreword in Dr. Rolf Schaffranke's *Ether-Technology,* Young Harris, GA.

can be gained in different ways. His theory explains how living organisms know and use information and that the role of information in nature is as fundamental as matter or energy. Importantly, his theory considers consciousness an essential part of the universe—the foundation of everything. He said, "Intuition is our first sense, not our sixth."

QUANTUM HOLOGRAM THEORY OF CONSCIOUSNESS

As have other physicists, Dr. Mitchell theorized that the universe is multidimensional and includes not only physical dimensions but also time—past, present and future. Although our physical senses and brain might perceive the world as solid, he proposed that it's more like a holographic image built up by interacting vibratory particle-waves.

His theory assumes that at the subatomic scale all material objects in the universe keep evidence of each event that's happened to them, and that the information is stored in a holographic form that can be retrieved by the mind when it "attends" to an object. His quantum hologram model describes all creation learning, self-correcting and evolving as a self-organizing, interconnected holistic system.

COSMIC STORAGE DEVICE

Rudy Schild, PhD, is an emeritus professor of astrophysics at Harvard University. He works at the Harvard–Smithsonian Center for Astrophysics. Science journals have published hundreds of his papers. Because of his long association with Dr. John Mack, he became interested in helping form a coherent understanding of the nature of space-time. As editor-in-chief of *Journal of Cosmology,* Dr. Schild tries to broaden scientific inquiry to include the nature of consciousness and what he calls the Universe of Universes.

"Conscious" now has broader meaning, thanks to Dr. Schild. If his hypotheses turn out to be correct, scientists will have to consider reality as consisting of embedded holograms, or a multiverse of realities which somehow give rise to the experiences we perceive.

His theory goes into uncharted territory. *Hidden Energy* won't attempt to explain his astrophysical observations; interested people can visit his website.[214] However, we believe his concepts make more sense than what we've been taught. For instance, our schooling gave us the impression that black holes suck everything in; we're told that not even light can escape. Yet Dr. Schild points out that they are the most luminous objects seen by astronomers. His theory about what he calls a "magnetospheric eternally collapsing object" is easier to imagine with Dr. Schild's analogy: "nature's hard drive that stores the vast quantum hologram information describing the entire past state of the Universe."

Combining his concepts with Dr. Mitchell's quantum hologram theory yields an interesting process. Quantum-hologram resonance information is stored in that cosmic storage device by way of the zero-point field. From there it somehow goes directly to a person's mind, which processes the information. The theories point to the possibility that science can explain how the universe could have cosmic memory, universal knowledge that could be tapped into.

We conclude that the door is open to a transformative worldview about the universe, because physicists are still figuring things out. Perhaps the cosmos is speaking through inventors to help humankind navigate through our greatest technological and environmental challenge—to shift from life-destroying technologies to life-enhancing technologies.

Dr. Schild and his colleagues are not the only distinguished scientists who have the courage to boldly explore consciousness. Others deserve a whole chapter about their work alone, such as William A. Tiller, Ph.D. His book *Science and Human Transformation: Subtle Energies, Intentionality and Consciousness* shows how to bring about beneficial changes in one's body, for instance. Dr. Tiller's scientific experiments reveal that our intentions affect the material world.

---

214   http://rudyschild.com.

CONTRASTING WORLDS CAN BE UNIFIED

Dr. Anita Goel has both a PhD in physics from Harvard University and a doctorate in medicine from Harvard and MIT. At a TED Med event[215] she talked publicly about happenings at the interface where physics, medicine and nanotechnology meet. She had started an organization, Nanobiosym, for working at that convergent point and creating new ways of solving global problems.

(A spinoff company shepherded her health-related invention GeneRADAR through the federal Food and Drug Administration and will take it to countries in the developing world.)

Does a high achiever at prestigious institutions take time away from business to ponder consciousness? Her TED Med talk revealed the answer.

Growing up in the racially divided Deep South of the USA in the early 1970s, she said, "I got to spend a lot of time meditating in nature with the chickens and peacocks in my backyard."

Her parents were young immigrants from India. She was born in Massachusetts and from age two spent her childhood in a small town in rural Mississippi. With all that spare time she developed a fascination with physics and mathematics as tools to understand the mysteries of the Universe. On the other hand, she was exposed to practical problems of biology and medicine. Her father, a surgeon, let her accompany him at a young age into the operating room. The family also traveled internationally.

Living at the juxtaposition of different worlds, she developed a deep intuitive conviction that "there has to be an underlying wholeness in nature, an underlying unifying framework that would bring together the worlds of physics and biomedicine."

"However, the deeper I went into study of both physics and medicine throughout my academic career at Harvard, Stanford and MIT, the more aware I became of how deep this divide is."

"Modern physics, developed primarily in the context of inanimate matter had not yet come to terms with life, living systems and stuff like consciousness,"

---

215   Goel, Anita PhD MD, Ted Med October, 2010.

"...modern medicine as we know it is practiced chiefly at the level of molecular biology and biochemistry. Very little is understood how physics—things like mechanical forces electromagnetic fields—play a role in biomolecular and cellular processes. So, I really wanted to find a way to bring these two worlds together."

"Most of the physics of the 20th century is developed around closed systems... Living systems are open. They are exchanging energy, matter and information with their environment and usually far from equilibrium. So, we need...to expand the language and the machinery of modern physics in order to come to terms adequately with life and living systems."

PART V
BRINGING IT INTO
OUR WORLD

CHAPTER 25
FINDING NEW WAYS TO SERVE

The problems of manifesting free energy technology in the public sphere are not really technological, but issues of integrity and sentience.

—Wade Frazier, Professional Accountant.

WE MEET YOUNG RESEARCHERS WHO EMBODY NEW WAYS to approach energy science, education, and a soul-satisfying life.

Plasma technician Andrew Theodore Murray, for instance, approaches the new energy scene with open-minded skepticism and the intention to be of service while keeping his own life in balance. He also brings a playful spirit and respect for nature's wisdom.

When Susan moderated a panel discussion at a New Living Expo in California, she introduced him as a sustainable energy research technician immersed in the fields of plasma dynamics, non-localities, material sciences, unified field physics, spherical harmonics, electro-culture, and radiant energy devices. She asked panelists what excites them enough to get moving in the morning. Murray replied that every day is another moment to interact with nature, since he and his research colleagues were developing inventions that model what they see in nature.

We persuaded him to tell his story and to suggest how his generation can bring a new culture into the breakthrough energy scene.

"A HUGE FUEL TANK IS RIDICULOUS"

In childhood, Andrew Murray had been one of those kids who asked questions all the time and would often be found drawing schematics for his latest invention. A memorable moment came when he was eleven years old, sitting at his desk inventing a type of amphibious craft to travel the world in, and thinking. "Wow. I'll need a lot of power to go on this trip. A huge fuel tank. This is ridiculous." He remembers halting that frustrating logic with, "Wait; this isn't fun. There's other ways to do the power supply."

Boisterous play was usually the way to spur his creativity. If he wasn't enjoying his hobby it wasn't worth the effort.

"So I imagined what could be possible. Through imagination and being in a fun state of mind, it dawned on me that 'Oh. There's all this *space.*' Once that hit, this wave of knowing came on. It said, 'Of course. Space is an actual power supply! There's so much of it, and this is how the universe operates.'"[216]

His next question was why humans still use hydrocarbons for energy. The complex answers—about greed, propaganda, and human resistance to change—would sink in as he became older.

Andrew's father was a satellite engineer for Lockheed Martin. His school teacher mother also supported Andrew's fascination with nature and technology. His parents told him "you can do anything you decide to do." Another influence was his joy in being a musician.

After graduating from high school early, in 2010, he earned an automotive degree from a community college at age seventeen. It amused him to be a teenager with job credentials.

He was intrigued by experiments with water as a fuel enhancer, but those interests took a detour when he was invited to join the Sustainable Living Road Show as an entertainer and activist. The young activists rode in two 40-foot biodiesel powered buses across the USA. Murray also walked for two weeks in a march to Washington, DC, to protest production of genetically modified organisms, and later helped the 2011 Occupy Wall Street movement as a handyman.

---

216   Jeane's 2017 telephone interview with Andrew Murray.

Science resurfaced in his life after a musician friend connected him with a plasma physics laboratory of a small research and development company, ThrivalTech. He moved to Ashland, Oregon, to be part of its crew. Its experiments ranged from attempts to make cars run on water to using duckweed from a pond to make biofuel. Local entrepreneurs who wanted green technologies provided a modest amount of funding.

It was a dream job for Murray—a chance to learn college-level chemistry, biology and physics while helping experienced specialists. The team's expertise included a chemist and a few electrical engineering consultants. Others had automotive backgrounds like Murray's, but in addition he had studied the breakthrough energy field since boyhood.

The ThrivalTech culture was what he calls friendship-based entrepreneurship. "You walked in and you hugged thirteen people before taking a seat… it was such an open community."

Andrew Murray (Chester Ptasinski photo)

He lived on about $100 per month as an intern for the first seven months, and cheerfully ate triple peanut butter and jam sandwiches for meals. He became what he calls a 'shop bum'—working and sleeping in the same building, coming up with new ideas while sweeping the floor and taking out the trash as well as while helping with experiments.

When a ThrivalTech proposal brought in money to hire another full-time worker, Andrew Murray was rewarded for his dedication and quick mind. He now had a paid job that he experienced as ideal. He enjoyed five years of research and development that included using forces within nature such as plasma lightning discharges. ThrivalTech developed patented plasma technologies for use with internal combustion engines to clean up polluting emissions.

A GENERATION HELPS ANOTHER GENERATION

After his job became a paid position, Murray gratefully volunteered on weekends by traveling at his own expense to homes and laboratories of some energy science pioneers to help them in whatever way was needed. He encouraged them to pass their knowledge on to newer generations.

One of the brilliant thinkers he visits is the mathematician Elizabeth Rauscher, PhD, in Arizona. With her irreverent sense of humor, Dr. Rauscher had in her past stirred up the playful as well as the intellectually rebellious side of colleagues at the Lawrence Berkeley National Laboratory in the 1970s. She played a key role in starting an informal Fundamental Fysiks Group of theoretical physicists in the San Francisco Bay Area. Some of them eventually became well-known by writing about consciousness, metaphysics and quantum physics. Her group was the subject of MIT professor David Kaiser's book *How the Hippies Saved Physics: Science, Counterculture, and the Quantum Revival.*[217]

In today's era, Murray created opportunities by taking initiative, asking how he could be of service, and stepping beyond his personal comfort zone. And he found that getting an older generation to participate with the young can result in accomplishments that benefit the world.

How can Millennials engage with influential people who can help create a new paradigm? Murray replied that trying to reach influencers—those people with high-level educational degrees or political positions—by going through a bureaucracy is inefficient. He experienced that "horrendous" bureaucratic process as being designed instead to support lobbyists or people who are already insiders.

The most direct line he has seen for Millennials is through connections that a family member or relative, however distant, might have. "Take a risk. Go out on a limb. Go to that person that your relative knows who works for the government or works at Lockheed, or who works at whatever place..."

In short, many in his generation have opportunities to get help from Baby Boomers (the demographic age group following the Silent Generation and

---

217   Kaiser, David, *How the Hippies Saved Physics: Science, Counterculture, and the Quantum Revival,* W. W. Norton & Company, 2012.

preceding Generation X) who are retiring from high-level positions, still want to make an impact, and care deeply about supporting the next generation and reducing the amount of harm caused in their generation.[218]

"There are a lot of amazing people who are compassionate and are in very high-ranking positions," Murray said. "Our generation comes in and goes 'what do you think about this project? What can help pursue this idea, whether it's people, resources, time or land?"

Murray shares techniques for dealing with everyday stress; his tips can refresh tired achievers of any age. If his muscles tense while testing a generator or he feels exhausted, Murray takes a deep breath and tries to see the challenge from a new perspective so that he can engage with it in a more enticing way. He sometimes makes up a game for himself about the challenge.

A playful approach might at first glance seem unimportant to a high-stakes venture. However, Murray meets some long-time experimenters, both academic and independent, who are frustrated to the point of embitterment by obstacles they face in securing funding, getting workshop or lab space, being ridiculed, in some cases virtually run out of the country. He observes those who are "getting slowly drained and forgetting about their own selves and what it takes to be a healthy human. You can lose yourself in being upset with this field."

Another psychological trap in their struggle is professional jealousy or a mindset of competition. "Remember that there are other people working in this field that are quite competent," he advises. "Just know your truth and know your gifts. Helping to bring out the gifts in other people is not going to hinder me from bringing mine out. We need everyone, their intuition, their uniqueness."

BEHIND CLOSED DOORS

Inventors are commonly frustrated when they learn that advanced energy technology has already been developed behind closed doors. Andrew Murray's

---

218 Baby Boomers, also known as Boomers, were born during a post-war increase in birthrates, especially in North America, from the later 1940s through the early 1960s.

grandfather worked at Lockheed Martin. He was the chemical engineering manager whose team focused on a dielectric material for the leading edge of the B-2 bomber. That's where high voltage would be applied to propel the bomber by using an electrical field.

"I brought it up to him that I was working on antigravity (electro-gravitic) technology.[219] I didn't know he'd worked on that. He looked at me and said, 'how do you know about this? That's top secret!' And he only talked very briefly about it. There was still this very strong cloud of secrecy, only four years ago. So a lot of this stuff seems to be already built, but not in the public sector."

Knowing how easy it is to become jaded and negative, Murray's choice is to take a deep breath and ask what he can do with his skillset do to help someone else with a different skillset to get something done.

Bridging between "fringe scientists" and academia will take much patience and effort, he notes, but will be worth it because they need each other. For instance, unorthodox scientists go out and study nature and build devices according to natural principles. Mainstream scientists have access to resources including sophisticated instrumentation for testing.

The approach of competing for the credit for a discovery must also be dropped, Murray said, and replaced by sharing of resources in order to accomplish what the world needs.

ThrivalTech eventually needed to prepare for product development, which requires different expertise than prototyping. Andrew took time during the restructuring to reassess his own life while spending time with family and in the community.

At the time of this writing, he was excited to be working in a venture named Vimanika that combines aerospace engineering with concepts of over-unity. (He refers to over-unity by the term acceptable to engineers: "coefficient of performance greater than one.") A ThrivalTech co-founder launched Vimanika with a nod to Vedic literature's descriptions of flying machines called vimanas.

Regenerating our world by shifting us to clean energy abundance is not up to one generation. We're all in this together.

---

219   The late Townsend Brown pioneered the field of electrogravitics.

MEANWHILE IN EUROPE

Regina Lamour was thinking about Millennials' role in an energy revolution when completing her studies at the University of St. Gallen, Switzerland, in 2017.

She had met a scientist who had worked for years on developing novel 'non-linear hydro-aerodynamics,' specifically vortex oscillations that can be applied in many technologies as Viktor Schauberger had done.[220] To her it was a new dimension in energy science.

After further research, she views breakthrough energy devices and their advanced technology as the shortcut to meeting the goals of the Paris climate agreement.

In her scholarly paper, *Sixth Wave of Innovation, Fourth Industrial Revolution, and Energy Breakthroughs as Levers of Paradigm Shift*, she wrote that many Millennials already take part in climate activism and will rise up and speak up "the farther we move into… higher risks and uncertainties."

Lamour cites surveys in Europe confirming that although many older adults are more frightened of terrorism, more young people see climate change as the greatest threats to life and democracy. More than 31,000 Millennials from 186 countries participated in one survey.[221] About 78 per cent said they are willing to change their lifestyle to protect the environment. Lamour wants to introduce them to energy breakthroughs as a path to take "without pushing our civilization and the environment to the point of irreversibility, without deploying risky geo-engineering and provoking new crises."

"The current situation is the greatest opportunity ever to grow spiritually, intellectually, consciously and culturally. And the quest for the social progress is an extremely powerful way to design a version of society where new values are supporting the energy breakthroughs."

She concluded her academic paper by saying the advanced-energy transition should be brought into debates on climate, policies, financial targets, and how humankind can break through in this century as one civilization.

---

220   Lamour, Regina, from her paper "Why conscious paradigm shift?" for a meeting convened by Adolf and Inge Schneider of Switzerland.
221   Global Shaper Survey "Global Outlook on Climate Change Perception," 2017.

CHAPTER 26
THE YIN OF NEW ENERGY

> *The yin and yang of physics, of life, of the galaxy;*
> *it's the combination that works.*
>
> —Leigh Richmond Donahue, author.[222]

YIN IS THE HUMAN ASPECT IN EITHER MEN OR WOMEN that protects and nurtures all life. A new way of moving forward with revolutionary energy technologies requires more inclusion of subtle and easy-to-overlook "yin" personal energy in a "yang" dominated field.

The concept of opposite qualities in male and female—yang as active and expansive, yin as receptive and contemplative—comes from traditional Chinese philosophy. In the natural world as well as in traditional teachings, forces that on the surface seem contrary to each other are often complementary, interconnected, and interdependent.

We use the yin-yang terminology as a discussion starter. Nature shows us that interplay, balance and diversity succeeds and maintains equilibrium in an ecosystem. The same could be said for a society.

---

222 Donahue, Leigh Richmond, *Field Effect: the Pi Phase of Physics*, The Centric Foundation, 1992, Lakemont, Georgia, P. 75.

YIN IS NOT WOMAN; YANG IS NOT MAN

Yang is the ancient term for representing masculine energy as rational, concerned with the physical, forceful, expansive and individualistic.[223] Both men and women need to have enough of it within their psyches to take action. On the other hand, traits classified as yin include intuitive, collaborative, nurturing, inclusive and focused on connecting.

Yin qualities are not limited to one sex, nor are yang qualities. For instance, concern about children and future generations is universal. Making those concerns and the emotional well-being of others a top priority is more likely to be observed in women. But not always. We know men with contemplative and other qualities that can be labeled as yin who meanwhile have wives who show excess yang by being competitive or pushing for quick action toward acquiring money.

British politician Margaret Thatcher exhibited strong yang regarding political power. The longest-serving prime minister of the United Kingdom of the 20th century, Thatcher became known as the Iron Lady.

Viktor Schauberger (chapter 7) observed opposite forces interacting beneficially in the landscape. For instance, the hydrological cycle alternates between dissipation of water into the atmosphere and its condensation into rain.

He spoke of polarity as "nature's engine." Harmonious interplay between polar opposites of repulsing and attracting forces at the atomic level are "the dance of creation."[224] He used both the opposite condensing/ inward-spiraling and expanding/ outward-radiating forces within the innovative energy devices he built.

Schauberger also had insights about interplay and balance in human society. He used the words masculine and feminine to describe yang and yin qualities whose imbalance for several thousand years resulted in too much aggression. He said nature shows that balance must be weighted slightly toward the feminine for creative growth toward a higher *quality* society. Otherwise, he said, quality degenerates, because the yang is more quantity-oriented in what it produces.

---

223   Bartholomew, Alick, *Hidden Nature*, Adventures Unlimited Press, Illinois, 2005, p. 52.
224   Ibid., p. 51.

Alick Bartholomew, author of *Hidden Nature: The Startling Insights of Viktor Schauberger*, reports that the natural law regarding qualities—about balance ideally being not 50/50 but instead slightly weighted—applies to all polarities such as matter and energy or matter and spirit; chaos and order; yang and yin; egoism and altruism; and quantity and quality.[225]

We appreciate the balance seen in inventors such as Walt Jenkins. They boldly take action, yet pause first to consider how the outcome might affect society or an ecosystem. Jenkins changed his approach to using seawater as a fuel when he realized its byproduct could harm sea life. To find a harmless technology, he temporarily halted that project in order to do further research, even though it slowed progress toward the marine fuel being commercialized.

A mindful path may indeed take more time. Being careful, attentive and heedful of consequences and responsibilities is worth it.

In the energy technology arena, men typically make the decisions of consequence. They are most often the financiers as well as the engineers or inventors who work on revolutionary devices. The conversation moves quickly to money and technology.

To allow for considering the human factor, the nuances of the interpersonal, and the 20-year-out plan, Susan encourages developers in the new energy arena to take the needed time. Instead of racing to create a product, methodology or project, they could first look at who is included or who benefits and whether the project supports something sustainable, and ask, "How is it supporting life energy?" In her experience, the yin looks at the finer details that too often are overlooked.

Lauren Evanow, inventor
and entrepreneur

Susan uses the term feminine energy more often than yin. She asked a few friends what, in their view, the new energy field might have lost by not better including it. Inventor/investor Lauren Evanow defined the term as

---

225   Ibid. p. 52

being collaborative, holistic and inclusive, and a quality of both men and women. Without it, nothing new can come into existence.

Society in general has lost balance and holistic thinking by not better including yin energy, Evanow said. "We have lost time in finding solutions to global issues. We will remain caught in our present global reality—which some describe as materialistic and destructive—if we do not include more of the feminine aspects of ourselves in everything we endeavor to do."

Isolationism and excessive rivalry have not worked, she said. Meeting face to face, deep listening, and working together is the way forward. "Educating women would be the most productive and positive thing we could do as a species if we truly wish to bring balance and harmony back into society."

A society seems to have become unbalanced when "tree hugger" is a term of derision. American environmentalist, economist and writer Winona LaDuke once said, "Someone needs to explain to me why wanting clean drinking water makes you an activist, and why proposing to destroy water with chemical warfare doesn't make a corporation a terrorist."[226]

Where is the balance when public funds support violent use of tear gas and water cannons (in below-freezing weather) against peaceful protesters? That happened in North Dakota where Indigenous people, the Oceti Sakowin,[227] protested against corporate plans to cross the Missouri River with an oil pipeline, knowing that pipelines can leak. "Security" dogs attacked the people when they walked onto grounds they consider sacred, which bulldozers had violated on behalf of Dakota Access Pipeline. In 2016, Standing Rock Sioux elder LaDonna Brave Bull Allard established a camp as a center for cultural preservation and spiritual resistance to the pipeline. That struggle for justice, ancestral homelands and clean water woke up people around the world who watched it unfold.

Extreme imbalance is seen, smelled and felt wherever industries dump pollution, all too often near where people of color live or near low-income communities. The 2017 study *Fumes Across the Fence-Line: The Health Impacts of Air Pollution from Oil and Gas Facilities on African American Communities* reported that nearly seven million African-Americans live near oil refineries.

---

226 http://www.honorearth.org is Winona LaDuke's website.

227 The proper name for the people known as the Sioux is Oceti Sakowin, meaning Seven Council Fires.

The industry's annual millions of tons of pollutants result in more than 138,000 asthma attacks per year among school-age children.[228]

Our point is that decisions regarding energy technologies have lacked what we might call vitamin Yin. For instance, a photo of delegation leaders at Paris Climate Change Conference 2015 shows a long lineup of dark suits, predominantly men. Their announcements did not significantly change business-as-usual.

On the other hand, Paul Babcock displayed balance of energies when he devoted much of a 2018 conference speech to the natural world and his concerns based in love of the life force. The deer, quail, rabbits, coyotes and swarms of bees that had previously frequented his hillside home are no longer abundant, apparently as a result of human activities. Without enough bees to pollinate blossoms, for the last two years his cherry tree produced only four cherries, and his neighbor's tree was just as barren.

In boardrooms where academic research funding is allocated, who considers the health of Earth's inhabitants seven generations from now? Who has the courage and commitment to stand against vested interests, forbid unwise geoengineering, and insist on a speedy transition to clean energy?

Emerging energy technologies introduced in *Hidden Energy* provide clean choices. One of our colleagues recently described another example. (Identities will not be revealed at this time.) "It could fit in your house. At half the size of a washing machine, it would give you more than enough power to heat hot water and provide you with electricity as well. The guy has been running his house with it for years."

Balancing the mix of decision-makers could help create a society in which all inventors feel safe enough to reveal their identities and location. In such a society, inventions labeled disruptive (because they disrupt corporate profits) would nevertheless be welcomed by our institutions if the inventions benefit life. Creating such a world requires careful and courageous planning.

---

228   The NAACP and Clean Air Task Force authored *Fumes Across the Fence-Line* in 2017.

CHAPTER 27
JOBS, HEALTH, AND
A TRANSITION

Artificial intelligence as it is currently conceived will very likely just speed up the mess....What we need, however, are solutions that come from our deep connections to this planet as beings of this planet, connections that no machine will ever fathom.

—Kurt Cobb, Washington DC based author.[229]

A FORMER DEFENSE MINISTER FOR CANADA, PAUL Hellyer, concludes in his recent book *Hope Restored*, "A super-fast transition to zero-point energy is the only hope of saving the planet as a hospitable human habitat and is by far the best way to produce a permanent robust economy for people who are unemployed or underemployed."[230]

Oilfield workers who may be displaced by the shift to clean energy sources particularly need hope, and young people in the Appalachian area of Kentucky know that their future probably won't be in coal mining.[231]

Even before a transition to advanced energy sources, plenty of jobs for them and others could be created in the field of well-known renewables alone, if there were serious investment in retraining and job creation. That would

---

229   Cobb, Kurt, http://resourceinsights.blogspot.com/2018/12/artificial-intelligence-and-limits-of.html.
230   Hellyer, Paul, *Hope Restored*, Credos, 2018, p. 177.
231   https://www.kystudentenvironmentalcoalition.org/just-transition-working-group.

obviously be a more life-affirming priority than today's massive government spending on surveillance, subsidizing fossil fuels, and secretive projects such as rumored space weaponry.

When former players in the military-industrial scene near the end of their lives, sometimes they admit the existence of those secret projects. For instance, Ben Rich had directed the Lockheed aerospace corporation's "skunkworks" (elite group working apart from other technicians on special projects.) Before he died, he revealed that, "We already have the means to travel among the stars, but these technologies are locked up in 'black projects'...and it would take an act of God to ever get them out to benefit humanity. Anything you can imagine, we already know how to do."

The spirit of helping others is alive and well, however, in numerous people who aren't waiting for governments to change priorities. The New Economy Coalition is one of many groups taking initiative.[232] It collaborates with other non-profit organizations, community-based businesses and cooperatives that work to bring opportunities to low-income communities, help villages and towns rebuild, and create new employment opportunities after resource-extracting corporations have moved on.

As this is written, American legislators are debating a proposed policy named after a historical forerunner. Former president of the USA Franklin D. Roosevelt's New Deal was a job creation project. During and after the economic depression of the 1930s, three million young men gained work experience in the Civilian Conservation Corps. They planted nearly three billion trees to help reforest America, built trails and lodges in parks, updated forest fire fighting methods, and built service buildings and roadways in remote areas.

Populations today concentrate in urban areas, so young people need different job training than in the Great Depression. An appropriate Green New Deal could train a workforce to build sustainable infrastructure, affordable housing and other projects while bringing economic opportunities to vulnerable communities.[233] However, in many regions massive tree planting is still needed to rebalance nature's cycles. In Canada where Jeane lives, corporations

232   https://neweconomy.net.
233   Kutz, Jessica, "The Green New Deal is already at work in one Portland neighborhood," *High Country News*, Feb. 1, 2018.

have increasingly pressured levels of government into lowering standards for forest stewardship. Excessive logging on hillsides results in mudslides and floods. Healthy forests benefit water cycles and climate and provide jobs as well as oxygen.

Early proponents of the Green New Deal in the U.S. envisioned tackling climate change while providing economic reform, and jobs in sectors such as renewable energy.

A GREEN REAL DEAL?

Most economists, politicians and statisticians have not yet acknowledged the reality of the ultimate renewable—the universal energy. Breakthrough inventions can open up even more types of employment when the reality of the universal energy is widely recognized. Eliminating expenses resulting from dependence on fossil fuels can allow national prosperity that can fund widespread job creation and retraining.

To bring new jobs to displaced workers, perhaps a Green Real Deal could supply long-term job opportunities to workers who had been employed in the fuel industries. (Those industries will continue at a smaller scale, selling petroleum-based products.)

Injustices could be alleviated at the same time. Indigenous peoples who have been targeted for defending the right to clean water could also have new sustainable economic opportunities. Prosperity resulting from clean-energy abundance could bring similar opportunities to communities of color. That would be justice, since energy industries often locate industrial pollution near the communities.

Changeovers to trustworthy clean energy systems would bring a wealth of benefits. Research from the Global Commission on the Economy and Climate concluded that bold climate action could bring a $26 trillion boost to the global economy between 2018 and 2030. The commission found that such action by business and governments could save more than 700,000 lives from air pollution, increase women's participation in the labor force, and deliver more than 65 million additional jobs by 2030.[234]

What's slowing the creation of those new nonpolluting-energy jobs?

The co-chair of that international commission, Paul Polman, gives one answer. He is CEO of a Dutch/British food and hygiene products supplier that reduces its environmental footprint by actions such as using less plastic packaging and favoring renewable energy. In an article for U.S. News & World Report,[235] he said that too often governments are a hindrance rather than a help, by continuing to subsidize fossil fuels. In 2015, subsidies handed to fossil fuel companies globally totaled $373 billion.

Handouts of that size don't go to renewable technology, and inventors of non-traditional clean energy devices are usually left empty-handed. *Hidden Energy* introduced a sampling of the many advanced energy inventions that offer non-traditional solutions. The efficiency of even simple prototypes in the new sector shows that they will dramatically outperform standard renewables such as solar and wind power.

As with other energy breakthroughs, manufacturing and installing LENR systems, for instance, would provide new jobs. Cheap heat would help industries to thrive and continue to employ workers. Devices such as those in the solid-state category could provide the electricity if LENR cannot generate it as efficiently as it heats buildings.

The variety and readiness of solutions to an energy crisis indicates that new megaprojects will be made obsolete by clean energy long before they can be completed. That includes fossil fuel, hydro dam and nuclear fusion megaprojects. Planners need to rethink many entrenched policies, including ideas for job creation.

---

234 2018 report of the Global Commission on the Economy and Climate.
235 Polman, Paul, "Action Needed to Tackle Climate Change," *U.S. News & World Report* Dec. 13, 2018.

CONVENTIONAL INNOVATION IS ALSO FLOURISHING

Renewables such as solar are improving and even overtaking carbon fuels in efficiency in some areas. Scientists are also experimenting with the material graphene, super-capacitors and advanced batteries for energy storage technologies.

Tidal power from strong flows in oceans off the coasts of continents has barely been exploited. Water is denser than wind, so tidal flows and waves can generate electricity more efficiently.

Waste heat, weight or motion are being reused via off-the-shelf technologies. For instance, the LucidPipe™ power system harvests previously-wasted energy from gravity fed municipal water pipes. The company is based in Portland, Oregon, and helps cities and towns whose water mains are large enough to have constant flow.

Jobs could be created worldwide by installing such systems wherever drinking water comes from elevated water towers or is piped from a reservoir in the hills.[236] Water flowing downhill builds up strong pressure that must be reduced before customers turn on their faucets, so cities install pressure-reducing valves. Instead, LucidPipe's spherical turbines insert into water mains. Flowing water spins the turbines and converts excess pressure into electricity.[237] Parts that need maintenance are located outside the pipe.

The largest cost savings, usually wasted, comes from insulating buildings. Insulation prevents heated or cooled air from leaking away and provides indoor comfort and security against freezing or against heat stress during power outages. Humankind could super-insulate nearly every new building, if political priorities were wise choices.

---

236 Thanks to Dr. Peter Lindemann for the **LucidPipe**™information.
237 https://lucidenergy.com.

ENERGY INTERTWINED WITH HEALTHCARE

Nikola Tesla's inventive genius went beyond his AC system of electricity and his later discovery of what seems to be a different quality of electricity. He also pioneered electro-therapeutic devices. Newer generations of such devices involving frequencies could replace many pharmaceuticals and some of today's invasive medical procedures. Resources such as Integrity Research Institute[238] can give you information about bioenergetic research, and the Consciousness and Healing Initiative (CHI)[239] has a newsletter and extensive network of scientists. CHI's aims to accelerate the science of consciousness and healing practices via transdisciplinary research, education and technology innovation.

The health enhancing inventions of Tesla and his successors in the field of vibrational or energy medicine are topics for a different book, although *Hidden Energy* did mention the Multiwave Oscillator and Dr. Randy Masters' research. Recognition of the life force is a common theme in alternative healing research.

Science fair exhibitor Titan Fraser, 8, and uncle Paul.
(Jodee Fraser photo.)

---

238   https://www.integrityresearchinstitute.org.
239   https://www.chi.is.

Many innovations aim at enhancing health, such as successors to 20th century inventions of French scientist Antoine Priore and American medical doctor Royal Raymond Rife.[240] Such devices are based on using frequencies to affect living cells. The late John Bedini advanced the understanding of Rife's device[241] as well as knowledge of what Bedini called free energy.

Children are eager to learn about a more hopeful future. Second-grader Titan Fraser of Kenmore, Washington state, and his sister Demaree won a blue ribbon at a 2018 school science fair with their display about "free energy," a topic introduced to them by their uncle. At the fair, the children handed out plans for building a tiny motor with a mechanical oscillator energizer originated by Bedini. He had said it charged its own battery from the universal 'Radiant Energy' described by Nikola Tesla.

Bedini had named his electronic circuit the Simplified School Girl (SSG) to taunt critics. They had not carefully followed his instructions for making it, so they failed to get positive results and therefore said it didn't work. However, a ten-year old girl had won a science fair in Idaho by following his instructions.[242] Titan's and Demaree's handout noted that Bedini also created a do-it-yourself Crystal Battery. Technicians keen to learn more about Bedini's work can buy DVDs from the *Energy from the Vacuum* series.[243]

It's difficult to judge which sector of national economies—healthcare or energy—holds more inertia and resistance to radical change. Multinationals profit hugely from pharmaceuticals, fuel and electric power. However, despite the political influence and financial wealth of vested interests, we are optimistic.

SUMMING UP

The need to get off fossil fuels is well known. Business as usual is a foolish choice, as is geoengineering the atmosphere with particles to reflect sunlight

---

240 https://www.theraphi.tech.

241 http://www.bedinirpxbook.com.

242 http://bedinisg.com.

243 The "Energy from the Vacuum" (TM) Documentary Series featuring John Bedini and Tom Bearden is produced by Energetic Productions LLC.

back into space. Adding more nuclear waste onto or in our planet is not wise either, even in the small amounts that thorium reactors would create. Fuel-less energy systems are safest.

Every day that passes allows more fracking, hydroelectric dam construction, and fuel spills to damage fragile ecosystems. Every day, more children develop asthma from breathing fossil fuel emissions. And extreme weather adds to the need for portable yet powerful clean energy generators.

Who will steer the shift to new energy technologies? Will they be controlled by governments representing the interests of weapon makers? By corporations that tear down rain forests, destroy soils and pollute waters in pursuit of quick profits? Or will the human family insist on sending its wisest thinkers and courageous stewards—rather than its top money-makers—to decision-making tables? Will it include those who care about the children of seven generations from now?

Energy revolutionaries don't have to confront vested interests head-on. A grassroots movement can quietly disperse information and garner support for developing the inventions that will help all the people of Earth.

Open forums on the other hand could be a mindful way for society to start planning a just and peaceful transition to new energy sources. Coming to agreement on priorities could be a first step, so we'll start the discussion by suggesting ten standards. Powerful new energy technologies should be:

- considered a sacred trust and used only for life-enhancing purposes;

- harmonious with the natural world and harming no ecosystem;

- mindfully selected along with plans for a peaceful transition;

- shepherded by the human family's wisest from many sectors;

- openly acknowledged and not kept secret;

- introduced to society in an accurate and reassuring manner;

- chosen and shepherded in processes that give voices of women equal weight;

- free from funding that could compromise ethics or result in weaponization

- justly distributed to benefit everyone, perhaps going first to where fuel industries caused the most damage;

- inclusive in regard to decision-making, with input from indigenous peoples and other stewards of land, water and human health.

You or your circle could create your own list. Consider making uncompromised statements of what you believe is best. Compromising with corporate interests for so many years has not served the common good.

Economist Milton Friedman blazed the downhill trail to today's extreme inequality of monetary wealth and destruction of ecosystems. He planted the notion that the main purpose of a business is to maximize profits for shareholders. That philosophy assigns lesser importance to concerns such as employees' health and how much time they can spend with their families, and the environment is often a low priority despite public relations hype. Respecting local control of a bioregion's resources does not seem to be a consideration, with Friedman's legacy. He won a Nobel prize in economics in 1976 and his views became the gospel in business schools. As a result, many aspects of our world are unsustainable.

However, if a sufficient number of people care enough to forge alliances, cross political lines, and unite for the common good, anything can be accomplished. A society that values living beings more than money and geopolitical power would mean better lives for hard-working breadwinners, and for Titan, Demaree, and all the children on Earth.

Even now, energy scarcity is a myth.

Charles Eisenstein writes in his book *The More Beautiful World Our Hearts Know Is Possible,* "Perhaps these technologies of abundance—of energy, health, time, and life—will leave the margins and take hold only when we, collectively, exemplify abundance ourselves through generosity, service, surrender, and trust."[244]

---

244 Eisenstein, Charles, *The More Beautiful World Our Hearts Know Is Possible,* North Atlantic Books, 2013.

CHAPTER 28
WHAT YOU CAN DO

Unless we have done transformational work on ourselves, we will remain products of the very civilization we seek to transform.

—Charles Eisenstein, author.[245]

THE AUTHOR OF NEW YORK TIMES BEST-SELLER *New Confessions of an Economic Hit Man* calls for people to support a new path for humankind, a life economy.[246]

According to John Perkins, when he was a young man who had served in the Peace Corps, he was seduced and recruited by secretive government agencies intertwined with what he calls the corporatocracy. His life slid into a different world. He became an "economic hit man," deceiving vulnerable leaders of countries into burying their citizens in debt so that they were forced to give up stewardship of their natural resources and other rights.

Eventually Perkins pulled himself out of that immoral trap despite threats to his life. He has for many years devoted his life to making amends. Months immersed in nature aided the healing of his spirit, and he founded non-profit organizations[247] that partner with indigenous people to protect ecosystems.

---

245 Ibid.
246 Perkins, John, *New Confessions of an Economic Hit Man*, Berrett-Koehler, Oakland CA, 2016.
247 Dream Change and The Pachamama Alliance.

A couple of years before Perkins' first book shocked readers, Jeane's friend told her that an unusual businessman had given a speech to the World Bank. The friend sent the text of the speech, highlighted where Perkins quoted an Amazonian elder: "Your oil and lumber companies and your cattle ranches tear down these forests. The world is as you dream it and your people have had a dream of big buildings, many cars and lots of industry. Now you realize that that dream is a nightmare."

Indigenous elders taught Perkins that since our world becomes what we dream it will be, we can re-dream what we don't like. We understand that to mean that positive visualization—of the world we do deeply desire—is a first step onto a mindful path toward a clean energy transformation. On the other hand, amplifying the meme of a dystopian future of scarcity and war only gives that possibility too much energy. Our creativity could instead be energizing a vision of civilization on Earth that inspires others into helping manifest it.

Perkins also told the World Bank audience that a decade earlier he had been escorted by two young Shuar men on a three-day hike to a sacred waterfall in the Amazon rainforest. On the way back, the young men stopped and bent down beside the trail, pointing at a tiny plant. "It was healthy three days ago..." That night the elders decided that the trail should be allowed to fade back into the forest because a tiny sick bush had spoken. The Shuar had listened. Perkins urged officials of the World Bank to learn from such mindful practices.

His further suggestions could guide any organization:

1. **Help future generations.** "In evaluating projects," Perkins advised, "...Remember that future generations of people are dependent on the wellbeing of future generations of plants and animals, all of nature."

2. **Protect every culture.**

3. **Ensure that all projects have a net positive environmental benefit.**

4. **Explore and feel relationships.** "...When you work in a country, relate to its people, to its natural world and the problems being faced by both."

5. **Take action in your community.**

When we interviewed John Perkins, Susan explained that many new energy researchers feel they have a mission to protect life on Earth. Some have inventions that, if optimized, could replace carbon fuels faster and cheaper than standard renewable energy technologies. Unfortunately, the vested interests that Perkins writes about influence academia, governments and the mass media. As a result, most people do not hear about "disruptive" inventions except in a negative manner.[248]

We wanted to know what Perkins has seen recently that gives him hope about a consciousness shift. What is different today?

Perkins was quiet for a moment before replying. Since Jeane had mentioned speaking to the International Women's Forum on Future Energy held in Kazakhstan the same summer that he had spoken at an economic forum there, he reviewed his 2017 and 2018 travels. He had found that people everywhere are looking for ways to heal our world and are ready for conversations that include the spiritual.

Wherever he went—Kazakhstan, Russia, China, India, Latin America—he met CEOs and many other people who were willing to learn. At a music festival in the Czech Republic, "55,000 people came for rock and roll, and 5,000 of them came to hear me speak in an auditorium about global economics and the need to change."

Perkins found that people are no longer shocked by his revelations about corruption. Instead, they ask "How do we change it?'

However, some in the West seem to forget that change begins at home, judging by vehement reactions to his blogposts about meeting with leaders and thinkers in certain countries.

WHAT YOU CAN DO—PERSON TO PERSON

Human unity requires a new mindset. Like Perkins, we have also been dismayed by hearing emotionally charged them-against-us messages. Susan is

---

248   See http://HiddenEnergy.org for a longer report on the interview with Perkins.

professionally trained in a skill called emotional intelligence (EI).[249] She sums up its practicality: "The high EI individual can better perceive emotions, use them in thought, understand their meanings, and manage emotions." Perhaps cultivating EI could help people work together to accomplish a shift as huge as transforming the energy paradigm.

Maggie Hanna, a geologist and consulting technical scout in Alberta's oil industry, has insights about bridging. Through her career she earned the tagline "Energy Future Girl." She helps oil industry insiders plan for transitioning to greener strategies.

Hanna's viewpoint is unique; friends say she lives at the intersection of Science and Spirit. Recently she suggested some ways to defuse conversations about politics, climate heating, privacy, gender, science and just about everything.

Maggie Hanna, Canadian technical consultant

Polarized opinions have become seen as normal, but they are not inevitable, she said. Human exchanges are generally rich, nuanced and complex. Hanna describes a "radical middle" approach that is far from a boring, silent, middle of the road stance. It means making space within oneself to really hear and understand what underlies another's position, especially when on the surface their words are the opposite of one's own treasured opinions.

The middle ground is between two statements that each may appear true and yet opposite. The tension between the two is the place of highest creativity and there is high value in being there, Hanna says, though it is rarely comfortable.

To explain the paradox, she notes that there are many levels of truth. "Small 't' truth is at the level of the story a person tells. It is true for them right now, and it can change in the next moment for no discernable reason or as they gain more information.

---

249   The term 'emotional intelligence' was originated by John D. Mayer, PhD, of the U. of New Hampshire and Peter Salovey PhD, now president of Yale U., in a scholarly article. Science journalist Daniel Goleman read it and wrote *Emotional Intelligence: Why It Can Matter More Than IQ*.

"Capital 'T' Truth is more universal, deeper, relates to fundamental qualities of the universe and of being a human being, and is true for everyone all the time. When we mistake one level of truth for the other, we can get into trouble."

Thinking "everyone should have my point of view because it's right" locks us into a narrow position. And if we avoid troublesome conversations with people whose thinking is different, we lose opportunities to deepen our understanding and even evolve.

Hanna told us that "It is not a *real* conversation unless we enter into it willing to be changed, or at least change our mind."

If we don't try to convert anyone, occupying that radical middle opens up the possibility for divided debaters to instead form partnerships.

THE MINDFUL PATH HAS REST STOPS

The first step to helping transform our world is to ensure that our own habits keep us in balance spiritually, mentally, emotionally and physically. Some scientists whom we meet have balancing techniques that also nurture their creativity:

Whenever chemical physicist Gary Schultz of California is struck with a realization about a technical challenge, he needs times of quiet afterward for processing the epiphany. "It percolates for a while, and having meditative time helps that process…. Soon after I left high school, my experiment became 'let me pursue being nice to my brain and see how it will be nice back to me.'"

You don't need to be working in molecular electronics, like he is, to benefit from his advice. Time for reflection is a gift available to anyone, even if it's only minutes per day. It means giving yourself an opportunity to slow down and hear the soft whisper of inspiration.

Schultz is concerned that the constant presence of electronic overstimulation is robbing people of that gift. He recommends information about that issue found in books such as *Disconnected* [250] and *Digital Invasion*[251] and in the Public Broadcasting System DVD *Digital Nation.*[252]

To access their own guidance, some people use kinesiology, a technique for testing the response of involuntary muscles when you ask yourself a clear choice question. They say it gives them answers to health questions, for instance.

Religious people benefit from contemplative prayer. Others tune in to what they call their inner guidance for help in everyday situations or crises.[253]

FUTURIST LOOKS TO INTUITION

Awareness of intuition is a form of personal hidden energy. Futurist John L. Petersen predicts it will be a survival skill. Petersen has insights resulting from extensive government and political experience. He founded the future-oriented Arlington Institute, and writes books about what he calls wild cards—high-impact surprises.[254] He also developed the world's first national surprise-anticipation system, for the Singapore government. Petersen's newsletter *FutureEdition* reports early indicators of potential changes.[255] He concludes we are in a time of major paradigm shift.

The confluence of changes in the world will continue to be so complex that people feel overwhelmed, Petersen told us. Increasingly, big changes are difficult to put into a familiar framework, yet if you close your eyes and ignore them you can be blindsided by an unexpected development.

The way to effectively deal with our times, he said, is with intuition rather than by just relying on logic. If you create a habit of accessing your intuition daily, it becomes second nature. You will then be able to trust your gut feeling when a seemingly incomprehensible situation requires a quick decision.

250 Kersting, Thomas, *Disconnected,* CreateSpace, 2016.

251 Hart, Dr. Archibald D. and Freid, Dr. Sylvia Hart, *Digital Invasion, Baker,* 2013.

252 https://www.pbs.org/wgbh/frontline/film/digitalnation.

253 Butcher, Anne Archer, *Inner Guidance,* APG Books, 2013.

254 John L. Petersen's most recent book is *A Vision for 2012: Planning for Extraordinary Change;* see http://www.arlingtoninstitute.org.

255 FUTUREdition is free from arlingtoninstitute.org.

THE NEW ENERGY MOVEMENT

We believe that a consciousness shift—a new paradigm of respect for all life—must arrive along with the energy revolution. More crucial than tapping into the ether, the most important meaning of *Hidden Energy* is the power source that we can tap into as easily as taking a deep breath and centering on what is most important. Basically it's the power of love. Countless people are recognizing the need for that consciousness shift, and many organizations work to help create a better world.

New Energy Movement (NEM) is a nonprofit organization based in the United States and started by physicist Brian O'Leary, PhD. He and NEM co-founder and environmental activist Alden Bryant, PhD, wanted to create a citizens' push for bringing in a new energy paradigm.

NEM was publicly launched at a forum in Oregon in 2004.[256] Afterward, Dr. O'Leary appointed industrial scientist Joel Garbon to be his successor.

Mark Hurwit, former secretary of NEM, described the organization's goals to potential volunteers. NEM aims to awaken people to the potential of New Energy—clean, sustainable and non-centralized power. Hurwit said the aim is to spread its message far and wide through "a citizen movement which will expand to express itself throughout the halls of political power, the academic and scientific communities, and in large and small locales...wherever we can make it clearly understood that not only would the development and wide-scale deployment of new energy technologies be a good thing, it is absolutely necessary.... It's a huge endeavor; the stakes are high, and we can use all the help we can get."

"The widespread adoption of all major scientific breakthroughs have required acceptance of a new way of understanding, and this is mainly what the New Energy Movement is about."

Susan is currently president of NEM. Recently she and John Cliss founded the Nui Foundation, registered in the United Kingdom as a community interest company. They describe the foundation as a living ecosystem for the development of moral technology. The two founders say its mission is

---

256 New Energy Movement conference "New Energy: The Courage to Change," Portland, Oregon, September 25-26, 2004.

to provide a credible central organization which connects highly innovative scientists, engineers, academics and other mission-oriented individuals with powerful problem-solving ideas. Its stated mission is also to connect "ethically competent politicians, financiers and business people with the resources to timely introduce Moral Technologies for the betterment of planet Earth and humankind."

Rudolf Steiner, the first influential philosopher to use the phrase "moral technology," described a future path for humankind's development where consciousness and morality would be an essential part of developing science and technology.

The Nui Foundation is set up to operate with balanced interaction between three "pillars"—its inspiration, legal and business hubs.[257] As this is written, plans for the New Energy Movement are to work closely with the foundation, focusing on public education and media strategies for the grassroots movement.[258]

HOW YOU CAN HELP THE BREAKTHROUGH ENERGY MOVEMENT

- Do your research, sometimes called due diligence, before forwarding information that promises free electricity. Scams harm the legitimate fund-raisers. If people selling plans for an electric generator hide their identity, keep your wallet closed.

- Be cautious when encountering hype such as "one device will save the world," or "we have already figured out Free Energy." A more collaborative spirit is needed.

- Be mindful that the change to clean, efficient, and potentially game-changing technologies isn't exclusively about the technology. It's also related to monetary systems and the market share which it can potentially disrupt. (Historically, people who have made

---

257 http://thenuifoundation.com.
258 www.newenergymovement.org.

money on oil, gas, renewables or nuclear energy are not inclined
to easily transition to a new way, even if it can potentially make an
enormous difference.)

- Know that every skill is needed in the new energy field, including
legal, creative, ethical, financial, leadership, science, manufacturing,
accounting, graphics, and nearly every skill you can imagine.

- Transformation involves education, activism, diligence and changing
of systems, so consider volunteering to help with the New Energy
Movement,[259] the Global Breakthrough Energy Movement (founded
in the Netherlands by Jeroen van Straaten),[260] or whatever life-
enhancing organization most interests you. You could organize a
local chapter.

- Dig deeper and learn more about any topics in *Hidden Energy*
that particularly interest you. Check the endnotes. Know that
the amount of information in this book (and added on http://
HiddenEnergy.org) is like the tip of an iceberg; more is waiting for
you to discover, explore and investigate.

- Be active locally. Work with your town, city, and other levels
of government to alert officials to different avenues that they
could pursue for clean energy sovereignty. Create a group in
your community to discuss the growing potential for that
energy sovereignty.

- Take initiative, after researching how to legally proceed. For instance,
a delegation in Denver proposed a city ordinance, "Protecting
Extraordinary Technology." It listed new energy as "zero-point,
over-unity, cold fusion, hydrogen production through water-splitting
using catalysts, radiant energy, permanent magnet powered motors,
implosion and vortex engines, and super-efficient electrolysis."[261]

---

259   http://NewEnergyMovement.org.
260   http://globalbem.com.
261   *Protecting Extraordinary Technology*, by Jeff Peckman submitted at the Denver city and
county building February 13, 2019..

- Know that you can make a difference by expressing your voice and using your area of expertise even if you are student.

- Share information—on social media, a blogpost, or in conversation. Make new energy a part of your vocabulary.

- If the information in this book is important to you or inspires you, ask your local library to purchase *Hidden Energy*.

- Envision what is possible with abundant, clean and free energy. How does life look for you in such a future? For your community? What problems could be reversed or solved?

- Know that helping can be fairly simple and natural. Start where you are, use your unique gifts, and do what you can without carrying a burden of worry about what remains to be done.

- Recruit helpers. Walk lightly on your part of the mindful path. Have fun. We on planet Earth are in a transition; be compassionate with yourself.

Isabella Dos Santos

We began Chapter One with the viewpoint of a university student in the Netherlands regarding energy breakthroughs. We conclude with words of Isabella Dos Santos, who is scheduled to graduate this year with a degree in environmental health from the University of Rochester in New York state. Isabella is the daughter of two Brazilian immigrants whose families came to America fleeing poverty, and she credits her upbringing for her values and ability to see the divine in all things. She contacted the New Energy Movement on her own initiative and now helps the organization as an intern.

After reading a draft of the *Hidden Energy* manuscript, she shared her thoughts, starting with the observation that fear has dominated our world and hierarchical class structure created a society of disconnected individuals. "To execute a transformative collective shift in consciousness requires courage, social intelligence and the overcoming of fear. The collective shift will stem from the courage to commit to something greater than ourselves; only then we will be able to overcome greed."

"The courage to challenge the limitations we have been taught will open the doors to allow imagination and truth-seeking to flourish. Courage is one of the driving forces that will counteract the forces mentioned by professor Lindemann."

She noted that ecological constraints affect human choices regarding whether to be selfish or altruistic, and "when the abundance of free energy becomes obvious, society will need more thinkers, more creators and collaborators. Fear will not be our principle driving force; connection and ingenuity will be."

"A fairer economic system will arise because of the decentralized nature of new energy. The power that has traditionally been concentrated in the hands of a few people will be redistributed to disfranchised and powerless communities resulting in their empowerment and freedom," she wrote.

Her dream is of a sustainable human society with justice, peace and healing at its core. If we and enough others energize that vision by infusing it with positive emotion, and take mindful steps toward it, a path will open up.

Your personal "hidden energy" can be as transformative to your world as any technology powered by the universal energy. In what small alcoves of your life can you replace anger with love? Our call to action is to *be the consciousness shift* as well as to support the clean energy transition in whatever way you can. As a first step, just imagine—in detail and with heartfelt gratitude—living in a harmonious civilization on a regenerated Earth.

ABOUT THE AUTHORS

JEANE MANNING is an author living in British Columbia whose previous books, starting with *The Coming Energy Revolution* (Avery Publishing NY 1996), have been published in ten languages.

Jeane Manning
(photo by Jessica Zais.)

Jeane was born in Alaska close to wildflowers and wilderness and loves the natural world. She also wanted to understand human nature, so in university she earned a Bachelor's degree in Sociology, with honors. Her work history ranges from social worker to newspaper and magazine staff writer and editor of a community newspaper. Meanwhile her priority was being mother to a daughter and two sons, now grown and with children of their own.

Thirty-five years ago, Jeane approached the new energy inventions with skepticism. Curiosity and the implications for cleaning up the planet drew her deeper into investigating. As a result, she wrote articles and books and has been invited speaker at energy technology conferences in five countries.

As well as communicating with readers via *HiddenEnergy.org*, Jeane will continue to be involved with sustainability projects and new energy innovation by working with the Avalon Alliance, a progressive group of companies based in Kelowna, BC. Avalon has a comprehensive vision for personal, workplace, community and world transformation. The organization offers sustainability service solutions, leading edge tech transfer, design-build management, green project training and wellness programs. Avalon's projects generally also involve municipal/regional planning and environmental/agri-lands stewardship.

Websites: ChangingPower.net; HiddenEnergy.org.

Susan Manewich
(Copperform Photography,
Asher Adams.)

SUSAN MANEWICH focuses on conscious leadership for the positive evolution of all life. She has spent over twenty years in the areas of leadership consulting, emotional intelligence, resonant technology and better understanding human dynamics to successfully transition through these planetary changes.

Susan works to bring ethics, integrity and global team cohesion to the field of New Energy Technologies (NET). She is currently president of the New Energy Movement, a 501C3 non-profit located in the US, and co-founder of Nui, the Foundation for Moral Technology which is a community interest company based in the UK.

Some of Susan's professional accomplishments are in the field of conscious leadership development and Emotional Intelligence, where her work has been consistently well praised by her clients around the globe including Harvard Business School, Yale University, University of Chicago GSB, London Business School and many others. Susan is a published author in several pieces of Emotional Intelligence work: she was a lead researcher and author with Dr. Jon Klimo on a paper entitled 'Scientific Information Received by Contact Experiencers' (2018) and is the co-author of *Hidden Energy* (2019). Susan has also given several public talks regarding consciousness and New Energy Technology.

Susan grew up in Massachusetts, USA, and was a participant with the PEER (Program for Extraordinary Experience Research) group in Cambridge, Massachusetts under John E. Mack M.D. She graduated with a Masters in science, focusing on organizational behavior and development.

Websites: www.NewEnergyMovement.org; www.HiddenEnergy.org.

EARLY REVIEWS FOR
HIDDEN ENERGY

"A seminal book that gives an overview of the exciting new developments in energy and transportation technology that currently are in the works. Jeane and Susan document the struggle and many human-interest stories of the heroes attempting to bring forth ideas that will completely change our lives for the better. They examine the political forces at work in an attempt to suppress competition from these amazing technologies which operate from a subtle energic source that pervades and animates the entire universe. *Hidden Energy* kick-starts a public dialogue by spotlighting the important questions we should be asking to bring this new paradigm into reality. Come learn and be part of an energy revolution that will transform the world."

—Paul LaViolette, PhD, physicist.

"Hidden Energy is an activist work for new energy answers. The authors spent over a quarter of a century culling through the nonsense in this controversial field to find the gems. Time is now on the side of the independent inventors where real creativity emerges from the next revolution in thinking. Energy is the key to the castle for solving many of the world's ills. A timely addition, Hidden Energy will move these technologies from the edge to the main street. Well done!"

—Dr. Nick Begich, Author/Science Reporter.

"In recommending this book, I feel that it gives some hope for the future and resolving energy crises. The areas of science, engineering and physics should be looked at in detail. Incorporating the spiritual nature of reality is extremely important."

—Elizabeth A. Rauscher, PhD, physicist and mathematician.

"Hidden Energy highlights innovators who challenge commonly-held belief systems in physics, medicine, and biology. Is there an unlimited sea of energy omnipresent in the universe waiting to be tapped? This book opens the door to more dialogue and raises questions that may determine the fate of humanity. Now is the time to assemble a diverse team of open-minded scientists, engineers, and students from around the world to discuss, research, invent, validate, and debate…Those who challenge the scientific consensus are historically those who propel humanity forward by leaps and bounds."

—Russell Witte, PhD, University of Arizona, Biomedical engineering.

"Hidden Energy is a tour de force of research about solutions to one of the biggest challenges facing humanity. The forward thinking inventors and scientists profiled in this book are not just iconoclasts; their work may be the basis for a clean, environmentally friendly future for all of us on this planet. This book was both eye-opening and thought-provoking, and I encourage all those who are interested in out-of-the-box science and technology to read and digest it."

—Rizwan Virk, founder of Play Labs @ MIT and author of *The Simulation Hypothesis.*

ACKNOWLEDGMENTS

For their support we especially thank primo copy editor Teresa Marshall, and all the Mannings and Marshalls whose love sustains Jeane, starting with Teresa and Rio, Jay and Jen, Stan and Kim, Nick and Sarah Marshall.

For the road travels, engineering expertise, professionalism and all things Steiner, John Cliss. And for the openness, joy and encouragement from Sidney Kornacki and the Manewich family, James, Bill, Kris, James Jr, Rachelle, Bill Jr, Nick, Jake. The Tierney's, Kris, Brad, Alysha, Abby, Shawn and Mark. Wordsmithing and patient support, Jefferey Jaxen.

Scientists corrected parts of the text, other friends proofread, and colleagues in the new energy field informed us. Those whom we thank include, but are not limited to, the following list. (Other researchers who contributed are already mentioned in the book index.)

Adolf and Inge Schneider, Alice Stevens, Andrew Michrowski, Anastasia Cooper, Betsy Finkelhoo, Callie Roang, Carmen Miller, Carol Hiltner, Catherine A. Fitts, Chris Holden, Chris Leaske, Christy L. Frazier, Claus Turtur, Dagmar Kuhn, Dawn Stranges, Dennis Briefer, Dianna Watson, Doug Kenyon, Ed Combs, Edgar Mitchell, Eli Call, Gary Vesperman, Guy Obolensky, Henry Curtis, Jackie Lindemann, Jackie Valone, James Robey, Jason Owens, Jim Dunn, Joe Martino, John Coelho, John Hutchison, John H. Reed, John Wong, Jordan Pease, Judy Gaudin-Reis, Karen Elkins, Karen McKenna, Kevin Dee, Laurel Zaseybida, Lauren Evanow, Mark Dansie, Mark Griebel, Mary-Sue Haliburton, Michael Mazzola, Paul Harris, Randy Ziesenis, Riel Marquardt, Remy Chevalier, Rolland Gregg, Rozjael Young, Rudy Schild, Sara Shannon, Sehbia Debria, Sepp Hasslberger, Sherry Blue Sky, Stefani Paulus, Steve Elswick, Steve Krivit, Terri Bernath, Terry Sisson,

Tony and Patricia Craddock, Travis Devlin, Tom Sparks, Travis Devlin, Valerie McIntyre, Vernon Roth, Vlad and Doina Plesa, and William H. Zebuhr.

If anyone who assisted us with this book is unacknowledged, know that the omission is unintentional and our appreciation of the help is very real. We also love the many souls who mentored us in the years before the writing of *Hidden Energy*.

INDEX

Printed in Canada